# Verboten Meteorology

By  John Billen

## Table of Contents

page
1.      Introduction
7.      Chapter1: Background and Theory
20.     Chapter2: The Cosmological Angle
60.     Chapter3: The Weather Modification Angle
83.     Chapter4: The Public Angle
118.    Chapter5: Benefits of the New,
        Costs of the Old
135.    Chapter6: Looking Ahead
144.    Chapter7: Anecdotal Evidence
156.    Chapter8: From the Beginning
169.    Chapter9: Suppositions and Predictions
180.    Conclusion

# Introduction

This presentation owes it's existence to the amazing advances in the natural sciences over the past 110 years, and beyond. For that I'm grateful beyond measure, to be able to live in a world where exciting discoveries are around every corner. Without the background knowledge acquired by reading about recent advances in science, this could not have gone far.

There is one discovery, however, that appears to have taken place around 110 years ago that very few people seem to realize even exists, and that being my only dissatisfaction with the scientific community at present, at least since the television show about the 4.4 million year old hominid fossil discovered in the highlands of Ethiopia,that is what I've felt compelled to write about. The main purpose when I write about this is gaining the attention of the world about the theory I have; having enough public opinion on my side to persuade the scientific community to give this theory an accurate and detailed analysis in the hopes of this theory emerging as truth, and finally, once that is actually accomplished, if indeed the truth is determined to fit the theory, then being able to look up "Weather Modification" in the Encyclopedia Brittanica and actually find it there.

I really want that, but I would also like to see the world a better place for all its inhabitants, and I think this theory I have would help a great deal in accomplishing that. The theories brought forth here, if eventually proven valid, could present the entire population of the world with a new alternative to passively accepting whatever weather comes along, good or bad. The local inhabitants of an area changes at national boundaries, and with weather systems being a natural event that respects no boundaries, there will be sure to be some overlapping effects that would spill over to neighboring areas. Hopefully everyone who reads this will see the necessity of having all the details discussed put in the proper perspective, since any attempt to alter the weather, at least from the viewpoint of the theory put forth here, would be a local phenomenon, confined to a few hundred square miles.

The first chapter will plunge right into the theory after some short discussion, followed by a chapter on cosmology and the search for dark energy and dark matter. The third chapter discusses the meteorological aspects of the theory if it proves consistent with truth. The fourth chapter is a bit of discussion about this whole concept in general and the impact it might have on the general public of the entire world. Then we get to some of the benefits, mostly, of this new device or devices and also the costs of letting this slip through our fingers. Chapter 6, Looking Ahead, does just that, but with the assumption that the theory is true.

Chapter 7 gives some anecdotal evidence, followed by what are really more like appendix chapters, chapter 8 is a look back to how and where I realized for the first time that weather modification was probably possible and is included for the sake of completeness, and the final chapter consists of suppositions and predictions that should be included, also as a kind of reference. After all, time marches on. Anyone who proposes a new theory in the natural sciences will give details and predictions of the theory. Will a theory stand the test of time, is the question. Only by making predictions and seeing those predictions borne out can the theory remain where others have fallen.

I really and sincerely hope that enough people read this and perhaps the meteorology community may come forth with good news,after a very thorough investigation of this, once public opinion is clearly leaning that way. I think the findings would be good news. I have tried to be as accurate with things as I could, and as you will find out soon enough, convinced there is something to all that is discussed that isn't immediately apparent.

A lot of the more recent discoveries in the natural sciences involve discoveries that are tested and true but are quite invisible to the naked eye, as in the case of the force of gravity. We know it is there, but there is no putting one's finger on it, so to speak. You can't point to it and say "this is gravity", or for that matter, any of the forces of nature. They exist, surely, and are constantly at work,but their workings are only induced after long and patient observation.

Black holes are concluded to exist, though no one has ever visited one nor is anyone likely to. They cannot be seen. They can be inferred to exist because of the tiny blank space at the center of the Milky Way, the gravitational effects observed there, and whatever other evidence. Dark Matter and Dark Energy are even more enigmatic than black holes and the forces of nature. But, indirectly, scientists have compiled enough evidence to induce that Dark Matter and Dark Energy exist, as well.

So there is quite a good little bit of natural science that doesn't meet the eye. Nevertheless, common sense can tell us a little about some of the things we can only speculate about. The universe has a certain regularity to it. Take, for example, black holes. They are former stars that have collapsed upon themselves gravitationally. What once occupied a huge area of space now occupies a substantially smaller space. We can make surmises as to what is contained within a black hole, and be reasonably sure that these guesses are pretty accurate.

In short, the interior of a black hole would have to be made up of a super hot plasma under tremendous pressure. The molecular bonds that exist between matter outside of black holes would no longer exist within one. Electrons, protons, whatever are all squeezed together so tightly that it is all one type of thing. We could be wrong, and some type of wormhole exists there, leading to somewhere else. That possibility is remote, however, though no one currently can definitively state that it is false.

If a star collapsed into a black hole, all the matter that was the star has disappeared. We have the two possibilities, and the hot plasma one is the more likely. Funny that what we speculate the interior of black holes consists of should mimic entirely what is speculated to be ejected from the Big Bang.

Consider also two different observations made by the scientific community. The first being that 96% of the universe is dark matter and dark energy, and the second the assertion that the vacuum of space contains energy. How confusing is that to you? If 96% of the universe is still undetected, surely no vacuum is a vacuum to an exact certainty. Whatever the dark matter and dark energy are, they may well be within that vacuum, undetected.

Beside that, the 96% dark components to the universe are broken down into 74% dark energy, and 22% dark matter, and that certainly doesn't make much sense either by itself. So some of the solutions to these problems are suggested by the theory I propose, an attempt to make sense of the data.

The weather may be the most discussed topic on the planet. Most of that discussion is almost part of the culture of many societies when greeting another person. To say "Hello, nice day today, don't you think?" can also be stated in many different languages. In this weather discussion we are going to go where few have gone before; to wherever the clues lead us and wherever imagination, intuition, and experience lead us from there, as deeply into weather phenomenon we can get with possibilities and theories. I hope the net result is even more weather discussion around the planet.

# Chapter 1.Background and Theory

---------------------------------------------
"Tom Thumb twiddled his
thumbs for 110 years."
JB
---------------------------------------------

Most of what you will be reading will be about electrical conductivity and resistance and weather modification, so we need to say a few things about the weather. First of all, we are going to assume throughout this book that everyone agrees that experiencing pleasant, livable weather is preferred over weather that could endanger lives. Naturally the goal of any attempt to modify the weather ought to have as its purpose making the weather less dangerous, and to make more water generally available. Somewhere, almost daily around the globe, some weather extremity can be found to be happening that everyone living in the area would much rather not have happening.

It should be noted that the stance of the meteorological community concerning modifying the weather is that apart from limited success with silver iodide cloud seeding, which increases cloud yields some 30% or thereabouts, no effective means of modifying the weather exists. There is a research project involving electromagnetic pulse propagation, the High-frequency Active Auroral Research Program, or HAARP for short, which is expected to eventually do things with the weather, but officially, not yet.

The Soviets have a complementary version of HAARP, called Squirrel. One can find accusations on the internet that the Russians use the equipment they have to create high pressure in the middle of the Pacific, leading to drought for the western U.S.. Chemtrails, exhaust from airplanes with special additives has been tried, as well. Accusations have been levied against the Canadian government on the internet for conducting such activity and not telling the public about it.

As for what one could hope to achieve if weather modification were a reality, growing more and healthier plants and animals on a world wide basis would be one benefit. No one starving to death would be a big plus. Never seeing water shortages or famine, knowing when it will rain with 100% certainty gives one a sense of well being that would be difficult to measure in monetary terms. Putting out forest fires with the new technology would also save lives and property. Making shipping lanes safer for ocean going vessels is another.

An absence of hurricanes, tornados, and floods would bring an end to the endless rebuilding of structures destroyed by weather, allowing many more buildings to stand for hundreds of years. Increased precipitation in the polar regions could help to build higher ice caps, and additionally drain the oceans of the strength to generate hurricanes. So is there really a choice for mankind should we find the magic wand that we could modify the weather with? We haven't reached the end of the benefits that weather control could provide and it seems from this point on that doing nothing with the technology, if it existed, would be far more dangerous than embracing it.

The atmosphere on our planet contains vast quantities of individual, extremely tiny things. The astounding smallness of these free floating gases is nearly impossible to grasp. That is what the science of meteorology investigates. Additionally, astrophysicists, cosmologists and others are currently maintaining that as much as 96% of all matter and energy is dark and as yet undetected. A concerted search has been mounted for dark matter and dark energy.

This implies that not only is the atmosphere comprised of tiny gases and a little dust, it is also comprised of particles even smaller in amounts as high as up to twenty four times the mass and energy of the atmosphere itself. These tiny undetected bits would most likely be precursors to hydrogen, and would be charged particles prone to electromagnetic influences. So meteorology gets a bit more complicated when you get right down to it.

When it comes to weather modification the thing that would make the biggest difference, and is most sought after, would be a means of creating precipitation. If there was such a thing, applying whatever it was would have a more or less to it, and a possible opposite in nature, so ending flooding might follow shortly after the discovery of how to make rain. Just a rainmaker would have uses too numerous to mention, rain equating with available water, and any and all uses that water can be put to by humans would have to be included,in addition to of making water more widely available to wildlife.

It looks as though a certain undocumented discovery was made involving electrical conductivity and resistance and weather modification. The first evidence of this appears with the experiments conducted by Nikola Tesla in 1899, in Colorado Springs with his Tesla Coils. A biography of Tesla[1] mentions that during the time when he was at Colorado Springs experimenting with his Tesla Coils on the platform above his laboratory an intense thunderstorm with some 12,000 discharges of lightning in a two hour period occurred once with all the lightning happening within 30 miles of Tesla's lab.

The book mentions that thunderstorms in the mountains are a common event in that area of Colorado,not far from Pike's Peak. The presence of such a large quantity of copper coils in a high place such as the Tesla Coils in Colorado Springs was perhaps the catalyst for the storm front that developed, by being large enough to create a path of least resistance for air molecules, dust, static electricity and possibly dark matter and dark energy to flow more freely along.

So Tesla tinkering with wireless experiments inadvertently showed to anyone who happened to notice a discovery that should have exploded onto the newspapers shortly after the third day, or maybe after a few weeks; that a large quantity of copper strategically located could change the weather. Communications being somewhat primitive at this time, something Tesla was seeking to improve, it could have taken some time for a newsworthy item to traverse the globe, and that apparently never happened. The electricity that Tesla pumped into the atmosphere found clouds to reside in and eventually be released as lightning. The clouds were probably there because the copper was where it was.

The magic wand was revealed for the first time, and shown to be rather too large for a normal sized physically fit man to wield. Aah, but mankind has forklifts and tractor trailers for wielding the rather bulky magic wand with which to modify the weather! Actually, mules and horses, pickup trucks, and additional manpower would likely come into play. After all, all one would really need to do would be position a quarter to half ton of metal somewhere high up, and the means to that end would be whatever was available or cheapest. In some cities in the Midwestern U.S., it could involve a long elevator ride, and a few carts.

Here, then is the theory, and the first human experiments that we know of where copper in quantity was placed on high, and remained there in place some time. Do you suppose the reference to Tesla having ideas about controlling weather could pertain to this, ideas that never got developed? From accounts of his trip to Colorado Springs his Tesla coils probably resided on the roof of his laboratory from early June 1899 until mid January 1900 when he returned to New York, possibly longer.

Let me give you a few parts from the same biography of Tesla;[2] "At dusk of that day Tesla had watched a dense mass of strongly charged clouds gathering in the west. Soon the usual violent storm broke loose which after spending much of its fury in the mountains,was driven away at great speed over the plains". A little later, it goes on to say that he felt he made a great discovery that day; "He

summed up the implications of this discovery thus:
"Impossible as it seemed, this planet, despite its
vast extent, behaved like a conductor of limited
dimensions. The tremendous significance of this
fact in the transmission of energy by my system had
already become quite clear to me. Not only was it
practicable to send telegraphic messages to any
distance without wires, as I recognized long ago,
but also to...(at this point as I read the thing I was
thinking the next thing must surely be about
weather modification)....impress upon the entire
globe the faint modulations of the human voice, far
more still, to transmit power, in unlimited amounts
to any terrestrial distance and almost without loss".
So he felt he had discovered how to transmit power
cheaply that day as well as send the human voice
around the globe.

One other excerpt does mention weather
modification;"I am positive in my conviction that
we can erect a plant of proper design in an arid
region,work it according to certain observations and
rules, and by its means draw from the ocean
unlimited amounts of water for irrigation and
power purposes. If I do not live to carry it out,
somebody else will, but I feel sure that I am right".
The book goes on to say that "This idea too went
into his legacy of unfinished business, and to this
day no one has implemented it".

Getting back to the two sentences "At dusk of that day Tesla had watched a dense mass of strongly charged clouds gathering in the west. Soon the usual violent storm broke loose which after spending much of its fury in the mountains, was driven away at great speed over the plains." The point is,the whole time that the Tesla Coils were in Colorado Springs, from mid-June of 1899 or before to at least mid-January 1900, there probably was not one consecutive week where no storms at all occurred, and were probably as frequent as twice a week, and sometimes in the mountains during periods of low pressure thunderstorms can threaten in the early to mid afternoon every day for weeks on end. All summer long, then, and into early fall, thunderstorms probably threatened almost every day.

Once the seasons change and temperatures fall, the storms would have become less frequent, down to around two per week. So there was the discovery, and if Colorado Springs had a newspaper back then the truth could be ascertained as to the frequency of storms in the area for that seven month period, and so the list of evidence grows, maybe. I'm certainly not going to take the time to try to find out on the internet, searching for something like that could prove difficult and time consuming. Hopefully, someone who works at the current Colorado Springs Newspaper will read this, and a more complete account of what took place will surface.

Finding the exact physical causes for what happens with a large quantity of copper on a mountaintop has proven elusive. The assertion maintained throughout this book will be that certain observable effects can be noted; these are falling barometric pressure, cloud accumulation, and precipitation,the latter usually occurring about 72 hours after placement. My contention is that what happened in Colorado Springs did not go unnoticed; further tests were conducted by persons unknown, resulting in the three decades of wet weather in the United States from 1900 to 1930.

The decade from 1930 to 1940 is now known as the Dust Bowl era, a time when experimenters unknown changed the experiment from utilizing copper, causing precipitation, to lead, which, being the classical non-conductor would have the opposite effect of copper, and tend to cause barometric pressure to rise and clouds to develop less readily. Actually, this dry period didn't last an entire decade, only around eight years or so.

Proving what caused this extreme change just as Americans were gearing up for even greater farming exploits after three decades of abundance brought on by increased precipitation along with new farming machinery being developed and mass produced at the time is probably impossible. I still think things happened then that everyone should have known about. The important thing to realize is that we don't need to know exactly what happened in the past, provided we have assimilated what knowledge there is to be assimilated from that time and are wise enough to use that knowledge here in the present and future.

As to how one can precisely pinpoint an exact cause and effect relationship between copper, or lead on a mountaintop or other high location and specific weather events, proof would only be inferred after numerous repeated experiments confirmed the same, or in the case of meteorological events, a similar result, in each instance. The evidence would need to be overwhelming to rule out coincidence.

Coincidence is very possible, one couldn't infer after one experiment that a cause and effect relationship existed between copper on a hillside and a storm that passed through after three days. One would be intrigued, and repeat the experiment. Since humanity is now much more well equipped to perform such experiments and accurately document the weather events that take place, I am confident that eventually my message will get across to enough interested persons such that someone actually does carry out such experiments, and additionally, report the truth of what occurred.

The fact that not all land areas on earth are mountainous should not prove to be an insurmountable problem when it comes to finding a suitable location to place copper where it would be one of the highest things around, and be capable of transmitting it's electromagnetic waves a long distance. In places such as the Midwestern United States where one can find only gentle hills for miles on end, the roofs of tall buildings would probably suffice to position the copper high enough to produce the desired effects.

Hence the long elevator ride mentioned earlier to position the copper advantageously. Solid objects would block any electromagnetic waves considerably, and decrease the effective range and strength of the waves, so there is the requirement that these copper placements be in an elevated location, with as few other land objects above them as possible.

The particular nature of this discovery is as much tied up with the physical properties of sub atomic particles, the province of the science of physics, or astrophysics, as it is with the science of meteorology, but that does not render any insurmountable difficulties to an aspiring student in one of those fields. A graduate student looking for an interesting subject upon which to prepare a thesis will eventually hit upon this subject and bring the matter to a scientific conclusion.

It sure is true that the years have piled up since 1899 when Tesla went to Colorado Springs doesn't make this assertion any more true or false. It only shows that the necessary work of actually ascertaining what happens hasn't been done yet, or if it has, the actual results went unreported. Nothing could improve the human condition more thoroughly than being in complete control of the weather at all times. A discovery such as this ranks right up with the discovery of fire, the wheel, the internal combustion engine, the computer, and other such important discoveries.

For successful control of the weather, cooperation on a global scale would be needed, and that shouldn't be seen as an insurmountable problem either. It could be that this discovery will be the catalyst for a change of attitude on the part of a large number of people toward other members of the species. If a problem needs to be resolved and the solution involves the cooperation of the entire population, such an undertaking might help to form more comfortable relations between the peoples of the entire world.

Water is precious, but it is also abundant in large quantities. The only problem is the relative distribution of the water over the land masses of the planet. Letting a metal do the work of a desalinization plant and more by placing it in a propitious location at strategic times could so cheaply redistribute fresh water around the planet and prevent extreme weather from destroying things that it is imperative that not much further time be wasted taking this idea forward.

Physicists have shown the atmosphere to be comprised of electrically neutral particles, not affected by the electromagnetic force of copper, leaving only the gravitational force to influence the tiny components of the atmosphere, a force that is 100 billion times weaker than the electromagnetic force. The stable isotopes of nitrogen and oxygen, $N_2$, and $O_2$, then, are neutral and that is 99% of the known air, and most of the remaining one percent is also neutral. So, a path of least resistance created by a quantity of copper on a mountaintop wouldn't appear at first glance to be likely to have much of an effect on the atmosphere, and consequently, the weather, at all. Nevertheless, there is too much compelling evidence indicating that changes do indeed occur involving copper and the atmosphere and other explanations need to be explored.

## Chapter 2. The Cosmological Angle

---------------------------------------------

" Everything is not now unbroken."
-- From A Song

---------------------------------------------

There is the likelihood that whatever fragments the Big Bang created have been condensing into hydrogen molecules all these billions of years and may still exist, comprising the missing energy and mass that astrophysics have been trying to discover. The smallest thing in the universe, the most abundant thing in the universe also, combines to create hydrogen atoms through some process we will probably never be able to see. Dark matter and dark energy are hypothesized because the universe is expanding at an increasing rate and given the total mass of stellar objects known, there is not enough mass to slow the expansion and eventually bring it to contract back upon itself gravitationally, creating another big bang.

So there must be more matter and energy prevalent that is undetected, according to current theorists. It looks as though it is not just theory. The orbits of the celestial bodies are taking place at a much higher speed than they would if what is observable were all that existed. Indeed, the calculations of the astronomers and astrophysicists indicate that 74% of the universe is dark energy,

22% is dark matter, and the remaining 4% is the matter already discovered. This may seem a little confusing at first since matter and energy are inextricably intertwined, but what it really points to is how much total energy and total mass are missing to satisfy astronomical calculations.

Actually, energy separates from matter when it is released, as in a nuclear fusion event. Energy is either released from matter, or is potential energy that could be released from matter. We would assume that the missing energy and mass are in the form of things that are a combination of both, not all energy being released from any current matter. In nuclear fusion, two hydrogen atoms release a lot of energy and fuse into one helium atom. This helium atom still has energy within it.

As to the percentage of dark energy that is already released from matter and the percentage that is potential energy, we don't really know. It is thought that all celestial bodies have a halo of dark matter and dark energy surrounding them, so we will conclude then that these halos of mysterious things are too small to detect by human sensing equipment, but truly must be there or orbits of celestial bodies would be taking place at slower speeds.

The two words matter and energy are arguably

two sides of the same coin. Actually distinguishing when something is energy and when it is matter is not exactly easy. Matter is used to describe something when one measures the gravitational effects of something. That same something, say a star, could explode in a supernova and what was matter and energy is now predominantly energy, so the coin has now flipped sides, although gravitational waves are still propagating from every atom, some of which are climbing the table of elements. Eventually, a new star with planets, asteroids, etc., with a smaller star at its core will coalesce, and the coin will have flipped back to some less energetic combination of matter and energy.

Energy refers to what is released from matter in nuclear fusion, in supernovae, even a pile of leaves on fire releases energy, heat. On the cosmological scale, energy also relates to the motions of things. The faster something is going, the more energy it has. So, the greater the energy, the greater the effect of seeming to pull the known universe apart. So, the elements, the matter that is known to us, can vary in how energetic they are. Matter is a word usually intended to mean mass, an object has so much mass, the table of elements gives atomic weights to each of the elements. Five pounds of something at rest has much less energy than five pounds of something moving near the speed of light.

In fact if it turns out to be one elementary particle from whence hydrogen originates it will then take the place of hydrogen as the most abundant thing in the universe, by far. A book by Dan Hopper, Dark Cosmos: In Search Of Our Universe's Missing Mass And Energy[3], has a part where, when discussing possible candidates for the missing particle he says it couldn't be a charged particle, because particles with charge would interact with photons, making them luminous, and thus detectable. But can he be absolutely sure that things in the world of particles too small to detect are happening as he thinks they are?

Some of the theories in unifying all the forces of nature may be beautiful and all, but an electromagnetic wave coursing through the air from a quantity of nearly pure copper may do something entirely unexpected to particles that small, and they might still remain non-luminous. So we would have to leave room for unexpected possibilities. A particle that is not charged, but behaves like one in the quantum mechanical universe, or a charged particle that is so minute that the interactions one would expect it to undergo when encountering a photon go undetected.

Magnetic monopoles have been hypothesized to perhaps have been created by the big bang, as numerous as protons, though it is assumed they would have huge amounts of energy. Some Grand

Unified Theories proposed that a transition took place early in the history of the Universe, in which the three forces of the Standard Model, the electromagnetic, strong and weak forces emerged from a single grand unified force. "As a consequence of this process, an enormous number of strange objects called magnetic monopoles would be generated."[4]

Magnetic monopoles appear to be the most likely candidate. These are magnetic particles with only one side of a traditional magnet. They are not found in nature in the sense that any magnets discovered or created have always been found to have two poles. This refers to objects large enough for humans to see and touch, and any magnet such as this when cleaved in two, will be two magnets, each with two poles. If the primordial specks were magnetic monopoles each one would be energetic, unstable, and spinning at varying speeds depending on what it comes close to.

Some unknown fraction of an actual hydrogen atom in size, they will never be detected, yet they snowball together, coalesce into hydrogen, and maybe in a million years we will be able to detect an increase in detectable matter throughout our known universe, besides finding more hydrogen in the atmosphere along a path of least resistance, due to the primordial specks being pulled toward it, and accumulating in staggering density along it.

 Magnetic monopoles meeting others of its kind
when the increased concentrations of these things
happens begins the snowballing effect until a
hydrogen atom emerges, seemingly from empty
space. The whole process of snowballing from
primordial specks to hydrogen might have twenty
different sequences of events where more specks
are added or it combines with a like chunk in the
process.

 What we have then, is an entourage of primordial
specks surrounding every star, planet, black hole,
asteroid, comet and whatever other objects of
mass are in the universe. Some of these primordial
specks may be in deep space, but most would be
gravitationally bound to some object or another. All
of the celestial bodies would also have some of this
astonishing amount of undetected
mass and energy within them, as the smallness of
these things would allow them to pass through
solid objects. In the very small world of individual
atoms space is far more abundant than individual
particles. With regard to black holes, the
primordial specks would not be so much around
them as within them.

 On the surface of the Earth, primordial specks
would be in great abundance, a veritable blizzard
of them. The extent they reach into outer space
around the Earth could be as far as hundreds of
miles. Within the molten core of the Earth, and in

the centers of other celestial bodies there is likely some hydrogen nucleo-synthesis ongoing, with the exception of black holes and some of the heavier dwarf stars where the pressure is too great.

As far as this pertains to the Earth, lighter elements within the molten interior of the Earth are continually trying to bob to the surface just as a balloon underwater, and volcanoes occur as a result. If more hydrogen is getting created the quantity of lighter elements and compounds increases, adding to the total of light objects being pushed out of the heavier molten metals, thus to some extent increasing the frequency of volcanoes. Hydrogen in the molten core would combine with oxygen, sulfur, chlorine, carbon and doubtless other lighter elements. Water, hydrogen sulfide, hydrochloric acid, and countless other compounds would result from hydrogen synthesis occurring within the Earth's core, and the new hydrogen atoms and whatever they happened to combine with proceeding toward the surface. Carbon dioxide, nitrous oxide and a few others contain no hydrogen, but most compounds do.

According to the prevailing theory of the origin of the Universe the Big Bang occurred some 13.6 billion years ago, and all matter began flying apart at an incredible rate. Eventually, the hot plasma, the scientists say, cooled sufficiently for hydrogen atoms to begin to condense, these collected in large quantities gravitationally to form stars, the stars began to burn hydrogen via thermonuclear fusion, creating helium, the next heavier element. I guess some stars proceed to burn mostly helium in thermonuclear fusion, and some even burn carbon, creating even heavier elements.

When a star burns up most of the fuel available to it, gravity causes it to collapse upon itself, then explode in what is called a supernova. These supernovae are the source of elements heavier than hydrogen and helium due to the tremendous pressure and heat generated in these events. It has never been explained what condenses into hydrogen, or if that process is finished now. All we get is that hydrogen condenses out of the plasma created after it has cooled sufficiently. Hydrogen is the starting point for all the other elements. The early universe was blown to smithereens, then, and the plasma the scientists refer to once cooled would be precursors to hydrogen, of an unknown constituency, primordial specks.

Since the initial fragments of our proposed hydrogen precursors each contain mass, albeit in an exceedingly small amount,the combined mass of all the fragments accumulating along the path of least resistance could exceed considerably the mass of the air itself. At the highest estimate of 96% of all matter being dark energy and dark mass, and assuming the Earth has around it a proportionate share of primordial specks, and increased concentrations along the path of least resistance increasing that quantity by a factor of two, for example, the dark matter and dark energy accompanying the atmosphere could be as much as forty eight times as much matter and energy as the atmosphere.

Therefore, these small entities would influence the air molecules gravitationally,air molecules that are not charged, and unaffected by the electromagnetic force of the copper. Besides which, any object not traveling along the path of least resistance with everything else would be continually bombarded by these tiny objects, though this may not be much of a factor because of the minuteness of the primordial specks. The smaller of the two things would be deflected.

The incredibly huge quantity of primordial specks accumulating along the path of least resistance would near a gas molecule quite frequently, and get deflected, but since these are charged particles that immediately respond to the electromagnetic force of the copper, the primordial specks resume course for the path of least resistance provided by the copper. So, are the stable isotopes of nitrogen and oxygen likely to sit idly by while the gravitational equivalent of Mt. Everest is cruising along in a specific direction all around them? No, they would join the school of primordial specks because of the law of gravity.

The newly formed hydrogen atoms that have begun to coalesce from the increased concentrations of primordial specks would quickly pair up as stable H2 isotopes, and rise, being the lightest element, even if now paired up. So, as these newly condensed paired hydrogen atoms rise, the moment one of these encounters a free oxygen molecule it joins with the oxygen molecule and becomes a water molecule, H2O. Ozone,O3, is continually being created and destroyed as oxygen atoms are lost or gained in the upper atmosphere by the stable oxygen isotope,O2. Clouds appearing where none were expected would be explained by this theory.

The H that is created by the increased concentrations of primordial specks paired up as H2 would rise to the ozone layer, encounter a free oxygen particle there and become a water molecule, and being then a heavier entity than the surrounding air molecules, promptly begin to sink back to Earth, and eventually become a part of the developing clouds. The combined influence of the mass of the primordial specks on the atmosphere and the additional H2O being created would more than account for the effects perceived.

There are other possible contributing causes for the effects perceived with copper creating a path of least resistance. The Jet Stream could change course as a result of a row of copper tubing changing the magnetic field of the Earth to some extent. It could be that the static electricity following the path of least resistance weighs more heavily than expected. Perhaps the correct answer to the question is "all of the above". Each little thing by itself is maybe not sufficient to cause weather changes by itself, but is a part of the equation. The combined effect of all the mentioned possible causes produces weather changes.

In my opinion, a large percentage of the effects observed are due to primordial specks and the astounding amount of mass they bring with them along a path of least resistance, in addition to the hydrogen that comes into being from the increased concentrations of these. It is clear though, that despite some uncertainty as to the exact causes of the increased clouds and precipitation, nevertheless a half ton of copper on an 8000 foot mountainside near an ocean left in place for 72 hours would provide evidence all by itself, especially if one chose to begin the experiment when no clouds or precipitation were expected in the next few days by meteorological forecasts.

The known universe is speeding up in it's expansion. The common sense reason for that is there is more matter beyond the known universe exerting a gravitational pull. A guess would give us a possible ratio as to how much dark energy and dark mass are local, and how much is matter and energy more distant if one took the totals we have currently.

74% is dark energy, which is the force trying to pull the universe as we know it apart, 22% is dark matter, holding our known universe together, and 4% is currently detected matter, so since matter and energy are both contained within matter initially, and we have 22% dark matter holding us together we probably have a proportionate share of somewhere around 22% dark energy trying to tear us apart in the local arena, and the other 52% dark energy would consist of the pull of more distant objects upon our ever expanding Big Bang, objects with matter and energy combined, just like in the known universe. That still leaves us with an 11 to 1 ratio between local dark matter and dark energy, and the known elements, neutrinos, etc.

Astronomer Vera Rubin spent a great deal of time measuring the speeds of orbiting stars. She published a paper describing her detailed observations of the motions of stars in the Andromeda Galaxy and her conclusion was that "for the stars to be moving with the velocities they had, there would have to be as much as ten times more mass in the galaxy than was visible".[5] Actually the difference between 10 to 1 and 11 to 1 is only around 1.5%. This figure surprised me as so small.

Other than those figures, there is the consideration that our known universe is still in a rather energetic state; possibly more dark energy is present here, and less attributable to distant massive objects. Lots of energy is being released in the stellar activity of 100 billion galaxies. Add to that the energy released in supernovae, and a good argument could be put forth that the gravitational pull of distant objects accounts for less than 52% of the total dark energy that we have yet to find.

But, stellar activity and supernovae are not dark energy at all, they are quite known to us, but one could impute more energy to the dark matter and dark energy, if one were to suppose them to be hydrogen precursors. Once hydrogen, they are then capable of participating in nuclear fusion. Nevertheless, something is causing galaxies to fly away from each other at increasing speeds, and to clump together in enormous clusters and leave huge empty places in the known universe. Besides that, the observations of Vera Rubin point to around a 10 to 1 ratio of primordial specks to known matter, and the proportionality guess gave us 11 to 1, so perhaps the figure for dark energy here in the known universe could in reality be 2 or 3 percent higher.

Almost all the theories in Cosmology start with the assumption that the known universe is the entire universe, that space/time is a fabric of some kind, and that there is something called vacuum energy out of which particles pop out. Could it be that a much simpler explanation might be the truth? Massive objects could still bend light waves even if space/time is nothing other than what they are.

Photons have been discovered to have no mass but they are definitely a something, which came from something with mass, a kind of energy, so why do we need a fabric? They are pulled by gravity. I cannot elaborate what the force is that is called gravity. Newton evaded the same question. How a force can reach through empty space is beyond me, yet common sense tells me that is what is happening. Space doesn't change, a force travels through space. It makes the most sense that it would be a wave traveling at the speed of light, which is the currently held view.

Particles don't pop out of vacuum energy, they synthesize from smaller ones. The recent findings that the vacuum contains energy by some scientists completely overlooks dark matter and dark energy, which may be responsible for the figures these scientists are arriving at. To conclude that an area of space is a perfect vacuum when the unresolved question of what and where dark matter and dark energy are seems to be looking blindly at the vacuum.

Anyone currently exploring vacuum properties should know well that a perfect vacuum cannot be assured, when anywhere from 90 to 96 percent of the known universe is still some undetected tiny object able to pass right through solid objects. If some energy is being detected, certainly the experimenters would have to consider that some particle or another was responsible for it. Any energy detected would more likely be from some unidentified particle than from the vacuum itself, after all, the only known representatives of matter that we know about, the elements, release energy.

It follows that whatever undetected particle or particles that truly exist would probably be capable of releasing energy either in their current state or once they become hydrogen, if they do become hydrogen. Once again, it is hardly necessary to attribute additional features to empty space; it is not known for releasing energy.

The more distant an object is from us, the faster it is receding from us because the more distant it is, the closer it is to undetected massive objects, hence the more distant objects in our known universe are more greatly influenced by the gravity of these other massive objects hitherto undetected. For the entire universe to have been contained within the Big Bang and now, after almost 14 billion years, the entire collection of 100 billion galaxies is not slowing down as it careens farther and farther away from the point where everything was at the Big Bang makes no sense at all in the framework of the natural laws as we know them.

Gravity would have begun to slow the expansion eventually, so now cosmologists are inventing repulsive gravity to explain the excess of dark energy. How about normal gravity, just elsewhere? We have neighbors on the fringes of the known universe. The mass they contain could round things out. Therefore, the Big Bang most likely didn't happen as theories now state.

If we throw away those three assumptions, and in the words of Henry David Thoreau, simplify, simplify, simplify, we perhaps answer these questions, at least as satisfactorily as we can given our limitations. From what I can gather there is a problem in physics concerning uniting the gravitational force with the Grand Unification Theory, or with relativity, or quantum mechanics, and perhaps these new insights into the probable location of the 52% dark energy, and also where the missing particle can be found to concentrate which makes up, in all probability, somewhere between 90 and 92% of all the matter and energy in our known universe, and is the primordial speck, can help the mathematical geniuses in the world solve that puzzle.

At present, more and more exotic theories are emerging to try to bring successful answers to what happened in the first few seconds of the Big Bang, why the known universe is now speeding up in its expansion, and where dark matter and dark energy are. The complexity and unusualness of these theories is an indication that some simplification is overdue.

The notion of space/time as some kind of a coexistent fabric came about originally as part of Einstein's Theory of Relativity. That theory and its predictions would all probably still hold true even if we reverted back to the old fashioned ideas of space and time. From the beginnings of language, events and things were perceived in the real world to take place and have what one would call duration. We now have many words to indicate various amounts of duration. Something obviously changes from when an egg is raw to when it is cooked, beside fire being applied to it.

The fire is applied only so long or the egg is burnt. Actually describing the passage of time is difficult without referring to it somehow, as in "The fire is applied only so long", the only so long referring to how much time passes. These same events, to be perceived in the first place, had dimension also, in effect they were large enough to be visible to human eyesight. So, philosophers mused over the duration of events and the dimension of events quite a bit way back when, and decided time and space were both immutable absolutes, eternal, unchanging, or something like that, others claimed they didn't really exist, so what if the new and the old idea of time and space are both wrong?

A space ship leaving Earth and traveling near the speed of light for 100 years upon then returning to Earth will find family and friends long dead, though only about five years or so would have elapsed to the inhabitants of the spacecraft. This tells us there is definitely a variable factor in Time. Time nonetheless appears to be ever moving forward, slower in the space craft than on Earth. Space is where all this takes place, the spaceship would have traveled an incredible distance, but could space alone still be an immutable absolute, in short the edges of the universe do not exist, for there is no way for space to not exist?

It would be difficult to imagine that there might be a far location inaccessible to navigation, if one had some device for instant teleportation. Space thought of this way could give us a clearer picture of things, since if the void stretches forever calculations change. If time alone ultimately admits of changes alone, together with some figures to calculate total hidden mass and energy, that may make some computer simulations come up with a more detailed picture of the universe.

There was a solar eclipse in 1919 where Einstein predicted that the light from stars that are behind the Sun from our vantage point would bend as they passed by near the Sun, and astronomers confirmed that this is what actually happened. This observation served as the basis for the space/time continuum idea.

Yet the crucial question we must answer is does space really need to undergo some kind of change for this event to take place? Could it be that Space, eternal, infinite in every direction, stays the same and the photons change course? The force of gravity still has not been clearly understood, and attributing a change to what must be eternal and absolute sends one in the wrong direction.

Whatever force field or wave gravity is, perhaps also lays beyond our instruments to comprehend other than as a quantum mechanical probability. What happens to the mass-less photon, considered to exhibit the properties of both a particle and a wave, could be simply owing to it's existing, being energy, even without mass. Photons are possibly being pulled by gravity, preserving the nobility of Space. No evidence proves conclusively that Space need change in any way, and besides, without knowing the Absolute Frame of Reference, which we will discuss more in chapter 8, we cannot tell individual identical sized sections of Space one from the other anyway, so how would we know?

That being said then, a new theory of gravity perhaps that fits all that has been discussed here will be developed, and Space being unchanged and eternal doesn't invalidate Einstein's theorems either, with some modification noted whereby space and time are two separate entities possibly, with only one, Time, truly admitting of variation.

Actually, a lot of physics uses space coordinates, and it is very convenient and expedient for equations that involve space and time to factor the two in together. I'm really not sure if that isn't what Einstein meant by space/time. It is a construct to facilitate one's understanding of the processes, but the actual reality is that space is perfectly flat and perfectly empty until something enters it, and it passively permits passage. Someone said only two people ever understood Einstein's Theory of Relativity, and the number isn't getting any bigger here.

Computer simulations of all this might even be able to come up with a very accurate 72 hour prediction of meteorological activity when one conducts an experiment with copper, and all the other figures we came up with for the primordial specks are inputted, as magnetic monopoles. Chaos Theory says it can't be done, but after all, if the force extends far enough, and is strong enough, and if the primordial speck is unstable enough, small enough, I think the computer could get it all right and give us an accurate prediction.

A computer could accurately gauge a quantity of primordial specks moving along with the atmosphere with 10 times more mass and energy than the accompanying N2 and O2, and also successfully estimate how much of an increase in concentration of these primordial specks would occur, given the weight and position of the copper. It would then, based upon how much mass had accumulated in the area, assign motion to the N2 and O2 particles based upon the gravitation of the total objects now present. So all the predictions made by the theory could maybe be seen to happen in a simulation where all the conjectural data of the theory has been added.

Some of the newer theories brought forth in cosmology include the possibility of other dimensions. Multiple universes have been hypothesized to exist right along with the universe we live in; we just don't perceive these other universes because they are in a different dimension. So, ten or eleven other Earths traveling along with the one we live on, where everything is almost the same as here, is the scenario that has been proposed, or at least that is what the amount of dark matter and dark energy suggests.

This multiple universe idea has gotten quite a bit of attention in the world of science fiction. A number of new movies and television programs have emerged with multiple universes as a central theme. "Sliders" was a TV serial that involved passage between our universe and other multiple universes. Sliders came out some years ago now, and is already done being produced. The puzzle of where the dark matter and dark energy are in our universe could be explained in this way, though it looks as though confirming the existence of such other dimensions and proving they exist may prove difficult, if not impossible.

An infinite number of other universes have been hypothesized also, though one would wonder how there could be more matter and energy than has been apparently interacting with our universe. If we took the estimate of 10 or 11 to 1 for the dark components to actually represent other dimensions then it would be stretching the imagination to consider that there were more than about 12 different dimensions, the other 11 beside our dimension being similar representations of the same universe, or perhaps anything.

This multiple dimension idea is about the only other feasible explanation for where all the missing energy and mass are besides the theory of primordial specks. As to which theory is correct, we will all have to wait for more evidence. The multiple dimension theory is weaker than the primordial speck theory, though, in the sense that other unexplained questions are explained by the primordial speck theory that are not explained by the multiple dimension theory.

The most notable difference is that the primordial speck theory explains where hydrogen originates, whereas with multiple dimensions that question is left up in the air. There is also the anecdotal evidence in chapter 7 that any conscientious investigator would have to consider. The rain falling where it shouldn't is indeed the most intriguing bit of evidence, and would almost have to happen if the primordial speck theory is correct.

What happens in the first moments of the Big Bang, a big question in astrophysics, was so tumultuous as to have torn asunder all the matter and energy to such an extent that all that was left was magnetic monopoles. These eventually condense into hydrogen, and that process is not finished presently. Somehow neutrinos come into existence also. The known universe contains approximately 100 billion galaxies plus ten times that in magnetic monopoles, primordial specks.

There is actually enough matter and energy to generate 1100 billion or so galaxies then, and the explosion that resulted when two super massive black holes collided at enormous speed while approaching each other gravitationally had to have been supernova like but larger by far. Matter that is condensed in a black hole is squeezed so closely together that the normal bonds between atoms is completely gone. There are no spaces between anything within a black hole, and the entire thing is under tremendous pressure. With a collision between two similar entities happens, the brightness of the explosion would exceed anything humans have ever seen.

Black holes will eventually be seen as the retrofitters of the universe. The second law of thermodynamics states that entropy within a system will always increase. The universe is always moving toward a more disorderly state. The only time or place where entropy doesn't increase, one could argue, is within a black hole. Within these enigmatic entities gravitational forces prevent almost everything from escaping, and the pressure is so great that molecular bonds are broken. They are a system unto themselves, arguably discrete and self perpetuating. Of course they exert a tremendous gravitational force upon anything else in the universe.

Within that system that exists separately from all else, every last little scrap of matter and energy that has been spread far and wide that comes within gravitational range of the black hole leaves the disorderly state that it was in and enters the black hole where it becomes part of the high pressure plasma within.

Every last little thing that comes within range of a black hole will be absorbed by it, and be converted into a part of the homogenous whole. A twenty billion year old light wave, incredibly faint, spread incredibly thinly, when absorbed by a black hole, changes into part of the plasma within the black hole. Entropy decreases, as a total quantity throughout the entire universe, when that happens. A forty billion year old chunk of lifeless matter, adrift in space at almost absolute zero temperature, when absorbed by a black hole, becomes part of the plasma within the black hole. Entropy decreases, as a total quantity throughout the entire universe, when that happens.

Once a black hole becomes large enough and collides with another, the two objects being the only objects of mass over a very large area, all the plasma is released in an explosion such as the Big Bang, and galaxies eventually form with energy in abundance, planets with liquid water, life, etc. So repeated cycles of systems of galaxies form, eventually burn themselves out, are absorbed by black holes, two black holes collide, and the cycle repeats.

Entropy increases throughout the time as energy and matter are thrown out of a black hole, until such time as that matter returns to a black hole and entropy decreases. A hot, pressurized, uniform plasma squeezed inside a small area is certainly a more orderly arrangement than objects of mass of varying sizes spread out over billions of light years and releasing energy throughout.

The centers of most if not all galaxies is thought to be a black hole. Some collisions must occur in the universe between black holes that aren't super massive enough to create billions of galaxies, given that in our known universe, there are nearly 100 billion black holes, one at the center of nearly all the galaxies.

The likelihood is that within one system of galaxies such as our own, galaxies orbit each other in clusters and when two galaxies merge, the black holes at the centers of the galaxies won't be approaching each other head on and would likely merge into a larger black hole at the center of a larger galaxy. The Milky Way and the Andromeda Galaxy are going to eventually combine, or collide, over millions of years. Black holes in our known universe after more billions of years will combine with more and more black holes and eventually become super massive black holes after all the stars have burned out and eventually there would only be a few left.

This scenario suggest itself; right now, we don't detect black holes huge enough to have spawned

100 billion galaxies, but that is certainly due to the black holes being unable to grow any larger than they are in the 13.6 billion years since the Big Bang. Actually, the Big Bang could have been more than one cataclysmic event over the last 14 billion years or so.

If, for example, a mini big bang occurs between two black holes on average every two billion years or so, a half dozen or so may have happened in our neighborhood, and if that happened within a larger universe, the gravitational pull of the rest of the universe would have all the observable matter from our perspective heading away from us at increasing speeds, just as we see now. So, there would really be no distinguishable differences between one big bang around 14 billion years ago and a half dozen of them spread out over the same number of years from our perspective.

The microwave background radiation might tell us more about that, but I don't know for sure. For instance, if a mini big bang occurred less than 3 billion years ago,the background radiation from the event would have waves that were less elongated than background radiation from much longer ago. But if all background radiation detected is all the same wavelength, that would rule out multiple big bangs unless we are only detecting the background radiation from the most recent mini big bang, and all the other previous ones are now too faint to detect at all.

100 billion galaxies combined with enough primordial specks to generate another trillion galaxies is quite a lot, so I'm guessing there have been more than one colossal explosion over the billions of years that our known universe has been around, in its current state. Maybe the most recent was 5 billion years ago, and that is all we now detect, and it was the fourth or fifth such event in the history of the known universe.

However it all happened, with dark matter and dark energy occupying such a high percentage of all that exists in the known universe one could argue that this suggests that perhaps not all matter was contained within one big bang, since if it were, more hydrogen would have surely evolved by now, that event having happened almost 14 billion years ago. The lack of developed hydrogen suggests that some of the known universe might have come along more recently, with less time for hydrogen to develop, and has conjoined with what was here before. The galaxies we can see are all from the first few events, and most of the dark matter and dark energy that is now intertwined with the galaxies came along more recently on the cosmic scale.

I see nothing to stop black holes in our known universe from continuing to grow, merging with other black holes and eventually becoming the only thing around for billions of light years, at which time such a super massive entity would pull, and be pulled toward, others of its kind. When that happens extremely huge black holes travel huge distances straight toward a like object, collide head on, and lo and behold, a Big Bang. What must be contained within a black hole fits exactly the material ejected in the Big Bang.

All we have to do is wait long enough, billions upon billions of years, for black holes to continue growing and devouring each other. Either they collide in something a bit smaller than the big bang, or continue growing to accommodate 100 billion or so galaxies between the two black holes that collide. Perhaps the human race will live long enough to witness a mini big bang, if those occur. More likely is that if such an event occurred now, it would be an extinction causing event.

Galaxies that are observable now are each being slowly absorbed by the black hole at the respective center of each. Since super clusters of galaxies orbit around each other, eventually the black holes will combine. In the end, three or four remaining super massive black holes head off in opposite directions, being already far apart in the expanding known universe, towards other black holes from other systems of galaxies.

Or possibly, after 300 billion years or so, there are only a few thousand black holes remaining to our known universe, and they all drift off toward other objects of mass elsewhere in the entire universe, to continue growing until they are so large that a significant area of space lies between them and a similar object toward which the black hole is pulled, resulting in considerable speeds and a huge impact.

Contrary to popular belief then, perpetual motion is possible, all you need is gravity. The fact that the known universe is, and is here now, confirms this. The simplest explanation for how the wilderness we call the universe exists is that it always has. No special contrivances are needed for perpetual motion to have always existed. If it is a true depiction of reality, that the universe is a system in perpetual motion, then it existing now means it existed in any past time.

The second law of thermodynamics does not hold within black holes, and these black holes serve to replenish the universe with new energy. The first law of thermodynamics, that energy is conserved within a system, holds at all times in the entire universe, no doubt. The entire universe probably hasn't changed an ounce in total weight ever.

Some explosion comparable to our big bang could have happened 20 billion years ago at a distance of

20,000,000,050 light years, and we will detect it in 50 years when the light from the event reaches us. Such an event is a possibility. Will astronomers be looking for distant and faint objects if they think the universe is all contained within the known universe? Is light even detectable from a distance of greater than 15 billion light years?

The most likely place to find objects of mass outside our known universe would be just beyond the most distant super clusters of galaxies. Unfortunately, that means that a great many other luminous objects lie between these hypothetical massive distant objects and our vantage point on Earth, making detection of these things difficult. In fact finding a black hole 15 billion light years distant, when such objects are completely non luminous, and thousands of galaxies lie between our vantage point and the black hole would have to be impossible.

Huge objects of mass outside our known universe but not that far away would be unlikely to be observed in an area of space that was relatively clear since these objects would have pulled things from our known universe towards them, hence the space between the distant objects and us would be cluttered with galaxies.

If more distant objects exist, and some of those are likely to be black holes, we should eventually expect to see among the most distant galaxies that one that was there has disappeared, and then another and so on as they are absorbed by a super massive black hole.

Unfortunately we probably have a long time to wait, since the events we are witnessing from the most distant galaxies occurred many billions of years ago, and the light has only now reached us. If a black hole is absorbing a galaxy visible to our telescopes right now at the extreme edge of the universe, we won't find out about it for another 14 billion years or so. The disappearance of an entire galaxy wouldn't take place overnight, either, the process would take millions of years.

With that in mind, other objects of mass outside our known universe would have to be either black holes, or some accumulation of galaxies similar to our own, possibly with some variation in size, since the universe is a wilderness, and gravity could pull two black holes together at any stage in the development of the black holes.

Right now we can only speculate about what happens to smaller black holes when they collide, there is likely a point of impact at some of these meetings, most would probably involve the two black holes orbiting each other until they combine. There is also the age of any accumulation of galaxies similar to our own to consider. There could be extremely old systems of galaxies that are no longer luminous, indeed, the energy is so depleted that the bonds between atoms have collapsed, and most of it is now black holes.

There is also the possibility that some of these systems of galaxies are comprised of antimatter. Our known universe could drift into a very old system of galaxies comprised of antimatter, and since that system is no longer luminous we would only find out about that happening by seeing galaxies at the edge of our known universe disappear as antimatter and matter begin to destroy each other on contact.

Here, in this possible scenario, are other unanswered questions. Is antimatter within a black hole any different from ordinary matter, since the molecular bonds are broken therein? I think that once the antimatter is a black hole it is the same as one that originated as ordinary matter. So an old system of antimatter, consisting of mostly black holes, would most likely behave in identical fashion to other ordinary black holes. Galaxies at the extreme edge of the known universe would just disappear slowly. Once again, these events would unfold over extremely long periods of time.

In a Cosmological sense, the observation that the lighter elements within the core of the Earth are continually forcing their way toward the surface that is a factor in the development of volcanoes has tremendous implications when one considers whether or not it is likely that there is other life in the universe. If supernovae create heavier elements, and planets get created at the same time, the super huge chunks of molten stellar material that eventually became Earth and Mars came from a Blue Giant or other type star in a supernova event,with a yellow dwarf star, our Sun, at the center of a Solar System with numerous planets with moons, asteroids, and comets emerging at the end of it all.

Mars being smaller than Earth, and farther away from the Sun, it would have had a solid surface much more quickly than Earth, and Volcanism had brought the lighter elements to the surface just like on Earth. So the scenario that gives us is Mars was probably very Earth like with liquid water on its surface maybe as far back as a few hundred million years before the first micro-organisms appeared on Earth.

Having liquid water, sulfur, carbon, nitrogen, oxygen, hydrogen, etc.,and the propensity of these lighter elements to grow as ever more complex accumulations of chains of molecular compounds brings us to right about where Earth was when life first appeared. On Mars, several amino acids could have developed just as here, and with one able to reproduce itself, it began to do so, or whatever, some polymer made the breakthrough of making copies of itself, and eventually whatever this reproducing entity was grows further in complexity until it could be said to be a living organism. I've seen the statement that amino acids are actually small engines in themselves, and each would have different mechanisms that a living thing with various parts could put to use.

With that consideration in mind then, the possibility that life developed at all on Mars is probably quite high; whether that life could attain the complexity of the multi-cellular organisms on Earth is difficult to say. Considering that Dinosaurs came and vanished on Earth and were huge, the colder climate on Mars, and the more rarefied air might have made it difficult for animals or plants to reach sizes where more complicated multi-cellular organization can come about.

The largest creature may have been some worm a half foot long, maybe, and now it is probably too cold there for life to hang on. Possibly for Mars, some multi-cellular organisms flourished in the first 100 million years after the surface solidified,and as conditions got colder on the planet, life was reduced to smaller forms.

Life developing on planets orbiting stars among the 100 billion galaxies, when one considers that almost all will have volcanic activity and lighter elements spewing out all over the surface some time after they are created in supernovae doesn't seem all that unlikely, then. Obviously, gas giant planets are a different situation, though one could conceive of a gas giant a bit closer to its mother sun where possibly liquid oceans exist that are incredibly huge when compared to the oceans here. With an enormous cloud blanket, a gas giant with liquid water and a warm average temperature would be a perfect haven for living things, except for eternal twilight because of cloud cover.

A lot might depend on what kind of orbit and axial spin it has, as for example, if a gas giant spun on its axis too rapidly and the enormous atmosphere was in a lot of turmoil, the surface of the planet would admit of harsh living conditions. Some with gentle atmospheric conditions probably exist though, possibly with 500 foot tall mushroom like plants that would dwarf the Giant Sequoia,and clever, large brained creatures with manipulative hands who dig holes in these plants for shelter.

Other assumptions can be made about living things on other planets. If the environment is a safe enough haven for multi-cellular life to take hold, it probably will. Once that has begun, survival of the fittest will leave creatures that survive only, so, since the development of sense organs, central nervous system and appendages developed here as the battle for survival raged, one would have to assume that those three attributes would probably be fairly common among creatures with some form of movement, on other planets.

A head and eyes and ears, then, could turn out to be a lot more commonplace than one might think. A mouth with taste buds, some kind of olfactory organ, all these kinds of developments could be predicated from the need to adapt and evolve, just like here.

The same things that course through our atmosphere, course through alien atmospheres. Light waves in various spectrums, sound waves of varying lengths and intensity, odors of numerous things, would be as abundant there as here. Eyes, ears, and nose had better get developed or said creature goes extinct. Eventually some creature succeeds and reproduces. After a million years numerous different species are found, all with the common ancestor that successfully developed the sense organs to survive at a much more primitive level.

In this hypothetical alien world it is now difficult to find an animal that does not have the survival equipment of various sense organs, a brain and nervous system, and four appendages. As here on Earth in the alien world a lot of the two appendages nearest the brain would have specialized adaptations to help the creature in possession of them to capture prey, break open seeds, and other manipulative activity.

Plants on other worlds would be things that are rooted or held in place underwater on rocks, and similarities here, too, could be inferred, roots and branches are the appendages of plants, and plant life elsewhere would also branch out in some way or another for nutrients. Naturally, it would be the differences and not the similarities that would be noticed more.

An intelligent alien could appear ghastly and foreign but actually have a number of close similarities to humans. Perhaps instead of a nose like ours theirs is a four inch wide slit between eyes and mouth with the eyes larger than ours and bulging out of the sockets some. Such a creature to us would appear hideous. Actually the capacity for the alien organism to process information about its environment, conceptualize, and remember details about past events could be nearly identical to us.

One could continue with logical necessity and the laws of physics and probably expand on all of the things that would probably prove inevitable, such as plants on other planets producing fruit that are edible for animals but also propagate the plant species and on and on. The DNA molecule probably has some very close relatives. It would not surprise me that someone has already written something about all the probabilities of alien life being similar to here, one could probably fill a book with all of the possibilities.

Naturally we are curious about life elsewhere in the universe, actually meeting other intelligent beings is another thing. The distances to other stars are huge, and the time involved in travel long and dangerous. It could happen, but even if it doesn't, we could probably rest easy in the realization that life would have gotten started on other planets as well, some of these would have eventually led to intelligent life, and so on.

Chances are an intelligent life form developed quite a long time ago. Recent estimates based on stellar evolution, and relative abundances of metals in stars far distant conclude that life on planets such as ours probably didn't become possible until the known universe had reached 10 billion years after the Big Bang. So if intelligent life managed to survive for millions of years they would no doubt be ahead of us in brain power and knowledge by quite a bit. They, too, would still have problems with interstellar distances.

Curious that if intelligent species developed, and this most likely has happened, they would have developed on a planet with liquid water, as we've shown, and liquid water on the surface there would have evaporated and joined the atmosphere, just as here, and one could say that these intelligent aliens also experience weather.

They, too, would have also noticed that copper, or anything that conducts electricity well and lead or some other nonconducting material in pure form of some specific quantity in an elevated location can cause changes in barometric pressure, and precipitation amounts, at some time in their development as intelligent beings. Weather could prove to be the most discussed topic in the entire known universe, not just Earth. It would be hard to take a survey, though.

## Chapter 3. The Weather Modification Angle.

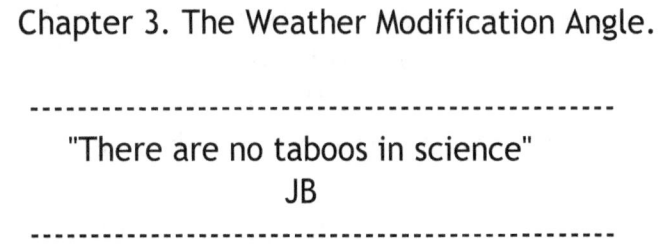

------------------------------------------------
"There are no taboos in science"
JB
------------------------------------------------

We mentioned earlier that Tesla had ideas about providing abundant water to arid regions, controlling flooding and such other weather control ideas. These ideas, and many others, were never developed by him. Those familiar with Tesla know that he was a prodigious inventor with numerous inventions and patents, and he was usually in financial trouble. A method of modifying the weather would not likely have yielded a weekly paycheck for him, whereas a number of other ideas he had offered a better chance of financial returns.

This should not be construed as any sort of criticism of the actions of one of humankind's greatest inventors. Tesla tore up his contract with George Westinghouse, throwing away his right to royalties on his alternating current invention, which paved the way for Westinghouse Electric to be able to afford to switch over to alternating current from the direct current in use at that time, championed by Edison, but far inferior to alternating current. So whatever his motivations and decisions were, we are still blessed by many things that Nikola Tesla invented and developed. Maybe we should add one more, and move on with this discovery. It has been claimed that an agency of the U.S. Government seized Tesla's private papers upon his demise.

Charles Hatfield came along in the early part of the 20th century, sometime between 1900 and 1920, and was known to have claimed to be able to make it rain. A meteorology text where I learned of this man stated that he was paid by the mayors of a number of towns and cities to work his mysterious cure for drought stricken areas. The book also mentions that he seemed to be successful in his attempts. The places where he was hired to do this work actually did see rain soon after he was hired and performed his duty.

What Charles Hatfield did was burn some copper compounds in a pyre type furnace, copper sulfide or sulfate, releasing plumes of copper compounds that rose into the atmosphere and joined the gases and dust that was already there and began to follow the prevailing westerlies just as the rest of the gases and dust do. He also claimed a number of different ingredients went into his concoction that he burned.

The author of the meteorological text I had perused regarded this Charles Hatfield as some sort of charlatan. In the opinion of the author these copper compounds or whatever Hatfield sent airborne didn't really have any effect on the weather. He went on to state that there have been similar claims about the presence of some metal or another having an effect on the weather, and concluded that these were all just instances of wishful thinking, not anything with a basis in fact.

That was that, apparently, all in one paragraph in a book with the title "Weather Modification", copyright 1980, some 400 pages in length. I never read the entire book, but I find myself wondering what the author could have possibly filled it with, since the only thing that changes the weather, according to meteorologists, is cloud seeding, and that only a little, and unreliably at that. Why, indeed, write a book about a subject that, according to the author and his peers, has so little subject matter? I looked up the title since I couldn't remember the author's name and found three books that could have been the one that I skimmed through and found that one paragraph and the mention of Blue Nile somewhere else in the book. These three possibilities are listed in the references, all copyright 1980.[6]

What little I know of chem trails, as they are called, is that some additive is added to jet fuel to make the exhaust different, or something is added to the exhaust as it exits the engine and depending on what those chemicals are, this kind of activity could prove effective in creating a path of least resistance. The similarity to this type of weather modification activity and the type of thing Charles Hatfield did are apparent. Hatfield, on the ground, maybe in a high location, introduced chemical compounds to the air in a furnace, with the rising heated air ferrying the compounds upwards to the height of clouds. The height of the jet that is releasing tainted exhaust could vary, but flying through the air at any height is a pretty good release point for a chemical that could prove effective in modifying the weather. So, in comparison, jet fuel seems more direct and perhaps more effective.

There is also, with both ground based and air delivered exhausts, whether from a furnace or a jet, a fantastic number of tiny particulates sent airborne that water molecules can latch on to and begin accumulating into raindrops, and if the molecules happen to also create a path of least resistance, there is the certainty that barometric pressure will fall, and clouds will begin developing. The developing clouds and the particulates ready to start raindrops might work together well.

However, one would have to consider the cost of being without whatever chemicals are burned each time since once they become exhaust the chances of recovering the material is non-existent. And, even if perhaps slightly less effective, copper placements could always be adjusted in amount to accommodate the creation of storms of the desired size, and, given safeguards against theft, always remain in the possession of the weather modifier.

The atmosphere is comprised of vast numbers of microscopic gases which have been fairly well described; about 78% nitrogen, 21% oxygen, a little argon and tiny amounts of carbon dioxide, methane and a few other gases. There is also water vapor in varying amounts, static electricity, sunlight, and dust, and possible primordial specks as before mentioned. All of the particles mentioned, not including static electricity or sunlight, are very small satellites in orbit around the Earth in the same way as the Moon is a satellite in orbit around the Earth.

The difference in size is quite pronounced, however, and that difference results in the orbits of all these very tiny satellites being very easily perturbed by all the neighboring satellites. These satellites that comprise the atmosphere are also much closer to the Earth than the Moon, so the uneven terrain of the Earth also factors into how easily and frequently these satellites are perturbed in their orbits.

Laws of inertia hold that an object in motion tends to stay in motion. This holds true even for the tiniest things. The fact that the orbits of these atmospheric objects are so easily altered doesn't change the fact that in the absence of anything causing them to alter course these objects would proceed in an orbit around the Earth similar to that of the Moon. The ease by which they can be made to alter course, then, is evident.

The magnetic field of the Earth varies at different locations around the globe such as the north magnetic pole and this can alter the orbits of our atmospheric satellites, something that doesn't happen with the Moon. The Moon itself is responsible, along with the Sun, for creating ocean tides, and the changing tides also influences some of the atmosphere's constituents. The Jet Stream has a huge impact on the direction of flow of the atmosphere.

Every moving object on land, or sea, or in the air changes some of the orbits of some of the atmosphere. I wave my hand, a whale breaches off the coast of Alaska, a jet plane flies form Los Angeles to New York, etc., all these movements of different things send atmospheric constituents scattering in various directions. As the Earth spins on it's axis, which is what gives us the 24 hour periods of night and day, atmospheric satellites are pulled along with the rotating Earth.

Hence we have winds that are known as the prevailing westerlies. These winds blow from west to east almost everywhere around the planet. At or near the north and south poles these winds are not as clearly defined because the top and bottom of our spinning planet doesn't travel as far in a day's revolution as the land areas away from the poles.

Currently it is held that gravitational and electromagnetic forces are a wave phenomenon. All matter exerts gravitational and electromagnetic forces. These two forces are invisible. They are both thought to be aspects of the single wave that all matter propagates at the speed of light equally in all directions. These forces weaken with distance. They are weak forces to begin with. Gravity is far weaker than electromagnetism, as we mentioned earlier. The gravitational field of the Earth is strong because the Earth is a huge object, and a man made flying craft must expend considerable energy to escape this field.

Copper is an element well known for its ability to conduct electricity. The reason that it conducts electricity so well is that copper atoms together in a piece of copper naturally arrange themselves in orderly rows. When an electrical current encounters a group of copper atoms, it passes between the atoms at the speed of light, encountering little friction in its passage. As with all matter, copper also propagates gravitational-electromagnetic waves at the speed of light.

Tiny air molecules, static electricity, dust,

primordial specks, and whatever else is airborne in the vicinity of a quantity of copper will encounter the neat, orderly arrangement that copper exhibits. All of these airborne objects would encounter less friction near the quantity of copper. A path of least resistance would exist in the vicinity of this quantity of copper, a path which tiny, free floating entities would flow more easily along than in any other direction.

This no doubt explains why Charles Hatfield appeared to have been successful in his efforts, and why Tesla's experiments saw intense thunderstorms. The copper compounds sent airborne, the copper Tesla coils on the platform in Colorado Springs, must each have created a path of least resistance along which everything airborne began to flow. Once the process began, the barometric pressure would begin to fall as a result of all the objects traveling in the same general direction, encountering each other less frequently since two or more things traveling in the same direction will not meet.

That is the essence of the matter, the fact that two or more things traveling in the same direction do not meet. Barometric pressure is in fact a measure of the degree to which objects in the air are encountering each other; a high barometric pressure reading indicates that the atmosphere at that point in time is undergoing more chaotic behavior, with interactions between air molecules happening frequently. A lower barometric pressure reading indicates that the air molecules are "schooling", moving in unison like groups of fish, large numbers of these atmospheric constituents traveling together in the same direction and interacting between themselves less frequently.

The dark matter and dark energy would be charged and definitely school as a result of a conductive metal of sufficient size being in an area, and the detectable N2 and O2 would have little choice but to follow a stream of things traveling in one specific direction many times more massive than itself.

How can I be sure the dark matter and dark energy are charged particles? Because the known particles that are electrically inert have gone through a number of changes, first becoming hydrogen, then helium, then eventually nitrogen or oxygen, then pairing up with a like atom before they reach the condition of being electrically inert.

Since the unknown particles are some fragment or another, looking to unite with others of its kind, they are unlikely to be so complete as to be inert and uninfluenced by the electromagnetic force of the copper in the hypothetical experiment. As we pointed out earlier, magnetic monopoles of such vanishing smallness as these primordial specks must be would satisfy the theory.

Since burning copper compounds results in a loss of scarce material, and since the same effect could be expected with placement of a quantity of copper, perhaps 500lbs. To 1000lbs., in as high a location as possible, that would be the most economical means to apply. My estimate is that somewhere from 500 to 1000lbs. would be a sufficient quantity, one that could also be retrieved and used over and over without loss of material. Some variation is to be expected at different locations around the world since some places are closer to an ocean than others. Precipitation will usually occur around seventy two hours after the placement of the copper on a mountain top, with a margin of error of 8 hours.

The surface area of hollow copper tubing being greater than solid bars of copper, hollow circular tubing would produce a more dramatic effect than bars of solid copper. If the copper is on a horizontal plane, a fairly robust wind will develop; if placed on an upward slope or vertically, there is less wind and more cumulonimbus type cloud development, and precipitation is more robust.

So the idea is to take coils of copper tubing, perhaps two feet in diameter, and arrange those in six foot tall circular things resembling a bird cage without a top or bottom. The diameter of the tubing itself should be one half to one inch,though a number of various dimension tubing could be tried. A number of six foot long straight bars of copper tubing could be used as supports,tied to the circular tubing with copper wire. Stand the six foot tall copper coil arrangements on end so they stand straight up,and since this will require a dozen or so of these contraptions,place them in a row from the west to the east, with a slight curve towards north, to assist in generating counter clockwise wind flow.

If you happen to be in the southern hemisphere, low pressure systems spin clockwise down there. Therefore a row of copper coil arrangements in the southern hemisphere would probably work best if it began from the west and proceeded to the east curving slightly south. After about two days the sky will be pretty much cloud covered in its entirety. From the second day on, precipitation could occur at anytime, depending on when night falls, but invariably will occur by the end of three days, perhaps 8 hours longer. If the copper is removed, the effect remains for a few days to a week with lessening intensity, the effect of inertia.

Often I have seen the observation made that rules that apply to ordinary things, or other disciplines in the natural sciences besides meteorology, don't apply to meteorology because the atmosphere is too complex on the Earth. It is true that it is a vast system with a number of unique land and water

features; The Grand Canyon, The Dead Sea, huge
deserts, rainforests, and oceans, to name a few.
The number of components in the atmosphere is
also staggering. No two rain storms are exactly
alike.

These considerations, however, are no cloak for
meteorologists to hide behind in a concerted effort
to avoid confronting this issue. If a predictable
quantity of rainfall occurs when a quantity of
copper is experimented with as I've described, and
if that quantity of copper being redeployed does
essentially the same thing a second time it doesn't
matter that there were differences in the rainfall
totals between adjacent counties in the area, or
that the center of the storm front was 20 miles
further north than the previous experiment, or that
one storm front moved due east, and the next one
took a slightly more northeasterly direction.

The results were still within the parameters of
the theory, and definitive conclusions could be
made, allowing for some variations between
individual storms. So suppose an experiment with
copper is tried when the current forecast calls for
no clouds for four days with a few high clouds
expected to roll in on the fifth day. A pair of
helicopters sweeps the area in advance for 500
miles in every direction for unusual barometric
readings, declares the area clear, and the copper is

placed in its location. A result consistent with the theory occurs, for example, rain began around 9 P.M., 68 hours after the start of the experiment.

A week later a location is found 600 miles distant with a forecast for cloudless skies for a week. The crew choppers to the new site, which is swept barometrically after transporting crew and copper, and within 6 hours of learning of the forecast have the new experiment started. 71 hours later precipitation begins, a similar sized storm system of the first one. This continues, how many times? What are the scientific rules to eliminate coincidence, or is that the decision of a panel of scientists?

Chaos theory predicts that if a system of things exceeds a certain level of complexity, predictions about expected events within that system are impossible. Yet if we can reliably predict the outcome of atmospheric events 72 hours into the future, this prediction holding true within certain parameters in every instance the experiment is tried, provided that is what eventually happens,we have rather proven otherwise, and have also proven that a cause and effect relationship does exist between a placement of copper on a mountain side, and an atmospheric event, precipitation, occurring around 72 hours after the start of the experiment.

Clouds block sunlight. Additional precipitation on

land means increased clouds over land. These clouds would be much more effective at helping reduce worldwide temperatures than the cloud generation over the oceans proposed by the Discovery Channel Project Earth program. The generation of additional clouds over land where things heat up far faster than water, things like concrete and asphalt, would be much more effective than cloud development out over the oceans where the additional clouds could contribute to hurricane development, especially since humanity has not yet begun to implement this proposed solution. Discovering, burying, and rediscovering this weather modification technique over and over will go on until eventually it becomes accepted fact.

Certain aspects of where an accumulation of copper tubing are resting can be important. A mountaintop, one would think, would be the best location. Actually, placing the copper a hundred feet or so from the summit of a mountain, on the west side, could prove much more effective in helping a storm front to develop. The difference in barometric pressure between the west side of the mountain where the copper rests and the east side which is effectively blocked from the electromagnetic transmission of the copper is perhaps pronounced enough to create a front line between the two air masses, facilitating the emergence of a storm front. A rooftop of a building with an additional tower added so that the roof of the entire building now has two levels could produce a similar effect if the taller wall of the new addition were to the east.

The question arises as to what happens to places on Earth that ordinarily receive ample amounts of rain when drier places begin to take in larger amounts of water. Will the Amazon or Indonesia or the Hawaiian Islands begin to see less rainfall? The answer is yes,the likelihood that some places could see less precipitation than normal is quite good, if human weather modification activities began in earnest. If the residents of these areas are aware of what is taking place and are aware that they are free to remedy the situation they are experiencing for themselves no real problem exists. There appears to be more than enough total water vapor available to satisfy every acre of land on the planet.

The oceans themselves would see less precipitation than normal if much more water vapor than usual were diverted toward land, so island chains would be affected. On the other hand, if the same residents are unaware that the weather is being modified in some areas of the world and are also unaware that they, too, could modify the weather, then there is a problem. When all residents of Earth can learn these things in an encyclopedia, there would be little doubt that should the need arise, appropriate steps could be taken to increase or decrease precipitation as needed, on a local basis.

Returning to the hypothesis that precursors to hydrogen still exist and are still churning out hydrogen atoms albeit at a slower rate than billions of years ago, do the activities of humans modifying the weather on our planet such as has been suggested here create more water than would ordinarily be created and could we be turning the Earth into a Water World, like the movie of a few years ago?

The answer to that question is probably no, since the hydrogen would be coming into existence anyway around the globe, and where a human act of weather modification was causing more hydrogen to be created there would at the same time be places around the globe where hydrogen synthesis would slow down as a result of the precursors to hydrogen having gone elsewhere, namely to follow the path of least resistance created by the human weather modification activities. The total hydrogen coming into existence on the entire Earth probably wouldn't change, so the total amount of water on the Earth would be unaffected. The Earth loses some of the atmosphere to outer space all the time, including water vapor, and the total quantity appears to be fairly stable.

Maybe after a very long time the Earth will have more water but if that happens it probably would have happened with or without human activity, and be a result of continued hydrogen production in the universe that we as weather modifying beings haven't changed, except to increase such activity in some areas while decreasing the same activity in other areas on the planet. To the extent that some hydrogen comes into existence due to the presence of a quantity of copper on the surface of the planet that would ordinarily come into being in the bowels of the Earth the act of modifying the weather could reduce the quantity of lighter compounds within the Earth, perhaps helping ease the frequency and severity of volcanoes. So, if that happens, slightly more water could be created than all the other compounds that would result from hydrogen nucleo-synthesis within the molten core. Whether that slight possible increase would prove significant over time is uncertain.

Crop yields would see an enormous boost from the lack of severe weather and the abundance of fresh water throughout the Earth. Certain plants prefer certain environments. The litchi tree requires a place with as little wind as possible since the litchi fruit are so easily dislodged by the wind, and such conditions could be made to happen for some areas. Rainfall amounts could be increased in spring and summer, and reduced in the fall for harvesting purposes. How many times has it happened that crops have matured and been destroyed by unwanted precipitation in September or October? This type of misfortune can be averted.

Thunder and lightning are usually more prevalent and likely to strike the ground when a quantity of copper tubing is placed in a vertical position on a mountaintop or building roof. The lightning could be captured by some copper scaffolding arranged so that when lightning struck it, the electricity would travel along the scaffolding down to an underground circle of super conducting material of considerable size where it could circle eternally. Hopefully, room temperature super conducting materials will be developed in the near future.

A few controls, some switches, and the power is transferred to the national electrical grid. The scaffolding would have to be no bigger than would cause medium sized storms in an area. An area that tolerates lots of rain well would be needed. A crew with a cherry picker to rebuild the scaffolding would be needed after damage from lightning strikes. A ship with the entire scaffolding and super conducting ring in the hold sturdy enough to withstand intense storms could prowl the ocean, or a rig similar to an oil rig could be constructed in a remote location in the ocean, far from frequently used shipping lanes.

If such an attempt to capture electricity is made, there is a possibility that if the copper scaffolding is large enough it may even be able to channel and collect ambient static electricity in the surrounding air and generate some power without lightning striking the device, though this would be a lot less than three million volts, which is the average voltage of a lightning blast. But there are small and big lightning bolts, so some probably exceed 5 million volts while smaller ones are only around a million volts or less.

How efficient would a scaffolding in need of repair from lightning strikes from storm to storm be? What percentage of a 3 million volt blast would get captured by the storage ring, and ultimately sent to the power grid? Questions I am unable to answer. I think it would be interesting to find out, but whether or not that will ever be done remains to be seen. Once in place the only maintenance necessary would be the repair to lightning damaged parts. A design consisting of identical sized replaceable parts could be used, and the damaged parts recycled.

Some insulation would be required for the underground part, or part of the hold of an oil rig type sort of experiment in remote seas, the crew would need a secure location for at sea and on ground activities like this, electricity captured out over the high seas would require an electrical cable along the floor of the sea, assuming humanity doesn't decide to transmit power around the globe as Tesla envisioned, and the impact on local climate would need to factor in with any land based such experiments.

What is clearly possible is that through examining this presentation eventually by enough people, and after some time for civilization to digest the new information, and new super conducting materials in production, perhaps a group of investors with a team of physicists and engineers might actually have a go at it. Again, computer simulations with the new information added would probably give an engineer or physicist a good prediction on events surrounding a rig all alone in the South Pacific with a significant quantity of almost perfectly conducting materials in the hold and also part of the mast and riggings above the deck, or the climactic changes expected with a try at it on land.

A steady condition of a sufficient quantity of copper being left in place indefinitely in the highest available location will yield a storm front every 3 to 4 days. Once the storm passes through a cold front with winds from the opposite direction of the storm will blow for a day or so followed by renewed cloud accumulation and the winds swinging back around to originate in the southwest, or northwest, depending on which hemisphere the experiment is in. This cold front that follows a three day event, after some hours of rain, is a very important cooling factor. The more times a placement of copper tubing produces a storm front over land, one can also count on the cooling effect of the cold front that follows, and all the clouds that develop that are themselves cooling influences over land during the daytime.

Leaving it in place may eventually  see a lessening in strength as nature accustoms itself to the new feature in the landscape; removing the copper for a week or so every two months or adding a few hundred pounds after a few months to bring precipitation yields back to the desired level might be needed. The sudden change is always more dramatic than something in permanent position. Winds will show a marked preference for the direction of a row of copper tubing. An acquaintance once said he saw an article in a mechanical magazine where they discussed using rows of copper around airports to assure that planes that are landing or taking off are facing a headwind rather than no wind or a tailwind.

By now the reader must be wondering, what about tornadoes? Up to now, nothing has been said about preventing them. As has been noted, there is considerable variation in the kinds of arrangements whereby copper tubing can be advantageously positioned; eventually enough different kinds of experiments will have been tried with each evaluated as to the effectiveness of preventing tornadoes.

For example, the litchi tree problem of needing as little wind as possible is probably solved by a slightly off vertical placement of copper, with a slight angle from the top to bottom part toward the west of about 20 degrees. Thus,the wind obeys the prevailing westerlies and at the same time the countering force of the copper slanted westward works in the other direction, causing a pronounced lull in the wind, lasting, no doubt, until the copper is removed.

As for how advantageously the farmers who are trying to grow litchis have positioned the farms, probably most litchi farming is done in valley areas. So the valley below the copper would see even less wind than the westward slanted copper in a higher location nearby. As for what arrangement would most likely bring an end to the "Tornado Alley" in the middle part of the United States each spring and summer, perhaps more rain in the desert southwest and large scale copper arrangements near the north pole each spring and summer would help.

Actually, the polar regions could serve as huge conduits for water vapor and static electricity on a year round basis. A permanent two ton row of copper on the northern Canadian steppe or Northern Alaska might just send enough weather phenomenon north toward the North Pole leaving correspondingly less in the middle of the U. S.. Polar conditions being as they are, the logistics of placing some copper tubing there in a high location may prove difficult, but not impossible. Once positioned somewhere exposed to the elements, it could be buried in snow and ice in three and a half days! So, a crew would have to visit occasionally and clear away snow and ice.

Usually what happens with tornadoes is huge storm cells develop out of thunderstorms, and these sudden buildups are hard to predict. A kind of copper arrangement could be tried where the angle was more to the horizontal to some adjustable extent with the wind, and cause winds to be just robust enough to keep a storm front moving without lumping into huge cells, but still yielding precipitation, and this might succeed in preventing tornadoes. Perhaps a multiple copper placement in the entire area, four perhaps spread several hundred miles apart, that together with helping more wind prevent tornadoes, also assured that individual storms are decentralized enough?

## Chapter 4. The Public Angle

-------------------------------------------
" I tossed them a silver spoon,
and they dropped it. "
JB
-------------------------------------------

There is a definite relationship between how much copper is placed near a mountaintop and how much precipitation occurs. One could deploy more copper than is necessary and cause flooding or tornadoes, possibly even hurricanes. Therefore, this discovery could be used as a weapon by anyone with less than honorable intentions. This could very well be the reason why no meteorological text contains any information about this discovery. It is simply too easy to cause changes in the weather, changes that could prove destructive in nature.

A ton of copper could leave three states underwater in about four days. Never should anyone attempt an experiment such as we've described with that much copper, unless they are in the Gobi desert and have already tried a half ton with little success, or are trying to extinguish a forest fire. This is the moral high ground that the scientific community, the media, and the government are clinging to desperately, about to be swept away by the tides of progress.

While in 1910 or thereabouts it may have been well and good to sweep this discovery under the rug, so to speak, and pretend that it doesn't exist, in these days of rapid technological advances, world-wide communications, and the current problem of global warming, this simply won't do. What do you tell your ten year old child when he/she discovers this on his/her own? What is the attitude of a group of teens toward authority figures after they have found this out on the world wide web? How does the current generation of adults appear to them, other than hopelessly inept?

It is apparent that the vast majority of meteorologists currently practicing that profession have probably never even heard of this possibility. Some meteorologists have in the past been aware of this, and chose not to communicate what they had discovered. The same holds true for people other than meteorologists. A few engineers surely learned this at some time, the early twentieth century was the age of the engineer.

Ordinarily, humankind leaves no stone unturned in the quest for truth, knowledge, and understanding. Yet here we are faced with a discovery awesome in the scope of its possible application for the betterment of the human condition. A discovery amazing for its stark simplicity, yet nowhere to be found in any encyclopedia.

The time is overdue to reassess the stance that has been taken so long ago towards this discovery and begin to formulate workable solutions to the threat of possible abuse of this feature of nature. No one wants to see the things they have worked all their lives for destroyed by the weather. A plan of defense is called for. Fortunately, there are things that can be done to minimize, possibly even eliminate completely, any deleterious effects that small minded people might try to inflict upon the population of an area.

Besides copper having an impact upon the flow of air molecules, a non-conducting element such as lead would have the opposite effect, logically. So, we are already in possession of a means to counter attempts to disrupt peaceful weather conditions. At the first sign of possibly severe weather lead could be placed in a high location so as to bring about higher air pressure, and consequently, less severe weather. In the United States there are countless barometric stations maintained all over the continent.

At the first sign of an unusual drop in the barometric pressure in a given area, one that is unplanned, a helicopter search for the offending material could be conducted, and using a barometer, readings could narrow the search area down to within a square mile or so, and since the material in search of would have to be in a high place to have any impact to begin with, I feel confident that finding and removing unauthorized attempts at modifying the weather could be accomplished within a few hours, far before the seventy two hours needed for a storm front to fully develop.

In the event that someone were to burn copper compounds after dark, Charles Hatfield style, lead might be the only recourse, though pinpointing the probable source of these copper compound plumes could also probably be accomplished, as for instance, when a substantial drop in barometric pressure sets in motion a search for a quantity of copper and no such quantity is found.

Wherever the barometric pressure readings were lowest would probably be fairly close to where the copper compounds went airborne. The area could be put under surveillance until the next time the culprits attempt such a misdeed. Ideally, only licensed meteorologists with the appropriate qualifications and permits would be legally conducting interventions into the weather in an area. Any other attempts by persons unknown could and should be considered an act of terrorism punishable by death. In contrast, consider the situation as it is now. Anyone with a pickup truck and access to enough copper (or lead) or the funds to pay for some could make life miserable for a lot of people.

A swift response coupled with a harsh penalty for such acts could make things far safer than they are now. Hurricane Rita and the aftermath featured a number of web sites that claimed this storm was a man made event, though there wasn't much elaboration as to how this was accomplished or by whom. There was a reference to electromagnetic pulse generation, such as the experiments conducted with H.A.A.R.P. So here we are totally unprepared for such acts in an era with such a thing as the world wide web.

A quantity of copper on a mountain top could be pinned down fairly accurately just by watching as the clouds develop. The area nearest the copper will have, in addition to some thicker clouds than the surrounding area, a misty, hazy cloud structure enveloping a one to five mile circle. Farther out, the clouds will turn into thin streams, all seeming to drift toward the area of thick clouds with the misty, hazy envelope.

To watch clouds one day, and see clumping clouds of the cumulo-nimbus type to the south, and clear skies to the north, with a flow of clouds welling up from south to north, if you watch the northernmost clouds you will see them slowly disappear before they fill the whole sky. That is a pretty good indication that lower barometric pressure is taking place to the south, and if one journeyed south to where one finds the thickest accumulation with accompanying misty, hazy cloud structure, that would be the best place to start looking.

Actually, nothing would be better than if this knowledge were simply public and the government did nothing to enact legislation, but simply let human nature run the course of discovery and familiarization, while the entire population of the world, ever alert to any possible dangerous developments, made sure disaster is averted. A neighborhood watch type of voluntary group would spring up, and the government need do nothing. However, I do not foresee that happening should this discovery ever reach an encyclopedia. The meteorological community would be elected to the position of sole weather modifier,most likely,after due deliberation by lawful authorities.

That is how the law stands now with regard to weather modification experiments, which most states have. If you want to modify the weather,you have to inform some branch of government in the state where you plan to attempt this experiment, and if you are not a meteorologist, it is unlikely that your request, or plan, will be approved by the authoritative body.

It seems paradoxical that the very group of scientists who have, up until now, left the public in the dark about recent meteorological advances will likely be shouldered with the responsibility of carrying out further extended use of such gadgets as can be devised. Knowing that this possibility exists, by the addition of this little feature of nature to the encyclopedia of the entire world, will at the very least alert residents of any area that unusual weather could happen, also that drought can be fairly easily remedied, and that they are not helpless to suffer whatever is thrown at them.

Lead atoms are quite unlike copper atoms. The arrangement of lead atoms features a very haphazard distribution, with no easy path for electrical current to follow. The friction encountered by electrical current makes it impossible for the current to pass through lead. Tiny gas molecules and other atmospheric entities find no easy path to travel along in the vicinity of lead, so the barometric pressure in the vicinity of a quantity of lead would tend to be higher since the gas molecules are encountering each other more frequently, having no clear path to follow.

Large quantities of copper could prove difficult to mitigate with the insertion of lead into an area; multiple tons of copper could prove extremely harmful to living things because of the flooding it could cause, and a mere ton or two of lead might prove insufficient to ease the situation. Fortunately, it would be difficult to hide such a large amount of copper in a high location, so the likelihood of it being found is pretty good. A barometric reading here, another there, and soon one could triangulate the probable location of the copper and remove it long before the intense storms it could cause have a chance to develop.

Naturally, too much lead put in place in a high location would lead to drought. This too, should be considered an act of terrorism and be strongly discouraged by the authorities. Barometric pressure readings should also give investigators a starting point in the search for outlaw lead placement as well, the difference being that the search would be centered on those areas with higher barometric pressure rather than lower. We can only hope that this whole idea doesn't escalate into a war between competing quantities of metals.

Right now we are helpless against any acts such as mentioned, as we have shown. Until the entire human race is able to learn of this discovery in an encyclopedia, the playing field will not be level, so to speak, and the risks are greatly increased with the situation as it is now. Human beings can adapt to changing conditions and new discoveries. I doubt that there would be many problems once the initial first steps are taken to begin a system of monitoring the weather that would include active participation in the making of, or breaking up of, storm fronts.

Given that this knowledge has been on the back burner for 100 years or so, the chances are pretty good that a number of people are already aware of this undocumented discovery. Since it is easily discoverable by persons independently of others who may also be aware of it, such as happened in my case, and since I have mentioned it to various people over the last 29 years,the number of people who are now aware of this feature of nature that has been described here has probably grown to the point where further concealment is probably impossible.

If you look up weather modification on the internet you will find that several companies already exist claiming to be able to assist with drought mitigation. The alleged techniques that these companies use varies. There is a lot of mention of electromagnetic pulse generation on the internet these days and it should be among the list of things that only licensed meteorologists should be doing, and some way of quickly and accurately finding such EMP broadcasts should they pose a threat to peaceful weather conditions should be explored. Of all the possibilities, electromagnetic pulse may prove the hardest to track down, but since it is also the most expensive the other methods would find more use.

Ideally, some further investigation into this phenomenon will be conducted, hopefully by meteorologists. The fact that any experiments conducted could be sabotaged by other persons with some other placement of metal in some nearby location needs to be considered. If secrecy as to the time and location of a proposed experiment must be maintained in order to assure the purity of the experiment, then the public won't know when such an experiment is conducted. Therefore, the public will have to trust the meteorologists involved in so far as faithfully recording what meteorological events take place.

For example, the meteorologists plan an experiment with 500 pounds of copper. An opponent of this idea becoming public knowledge learns of this proposed experiment, and places a ton of copper on an adjacent mountaintop, or building roof, as the case may be. Three days go by, and the resulting storms are now too severe. The meteorologists conducting the experiment decide not to release the results of the experiment they conducted.

Or, better still, the meteorologists plan an experiment with 500 pounds of copper. An opponent of this idea becoming public knowledge learns of this proposed experiment, and places a ton of lead sheets across a nearby mountaintop or building roof. Three days go by, and little or no precipitation occurs. The meteorologists then report that the experiment failed.

It would be imperative that these experiments be conducted with as much secrecy as needed to reduce the likelihood of sabotage. Only then would the true cause and effect relationship reveal itself to observers. Whether those conducting the experiment are answerable to anyone and whether the results will be reported honestly and accurately during these experiments one can only hope. Naturally results of these experiments wouldn't be learned by the public until some time after the experiment happened, which would be fine since the experiment was secretly conducted.

In the final analysis it may prove more difficult than one would expect to validate what has been asserted here by properly professional people in a properly professional setting. Obviously, if anyone over the years had happened to notice that this little feature of nature did what it did, they did not communicate this to the general public, or were prevented from doing so by a paranoid and overly militaristic government.

I don't know why this feature of nature isn't already in the encyclopedia. Many objections were raised by persons with whom I've spoken to the effect that if such a thing were true we would already know about it, therefore it can't be true. That is hardly a satisfactory conclusion either. The possibility that the truth is still to be discovered with regard to this phenomenon still holds true, even if it would seem to most that such a thing would already have been discovered. To assume that HAARP will be effective one day as a device to modify the weather with, one would also have to assume that the electromagnetic waves that propagate from purified metals would also have some effect.

The machinery creating electromagnetic pulses would also broadcast waves in the electromagnetic spectrum. So depending on the wavelength, high or low pressure could be created by these pulses. The HAARP experiments involve firing concentrated pulses at parts of the ionosphere, not just sending out random pulses at some wavelength, though it could probably do that as well. How much expense is electromagnetic pulse compared to placing purified metals in a high location?

Electromagnetic pulse would no doubt be overwhelmingly more costly, but the use of it could have military applications such as disrupting enemy electronic equipment, and I'm sure a few others,to the most extreme account of its possibilities, zapping someone off the sidewalk with it, so experiments continue.

Concerning the impact of this discovery upon world markets, it probably wouldn't be severe enough to disrupt agricultural futures. After all,the mere fact that Humanity has discovered one more thing to help them survive happens all the time. Truth is, agriculture won't suddenly become the easiest thing to do on Earth. There will still be work involved in growing things, caring for them as they grow, and bringing them to market after harvesting. Pests that threaten certain crops could proliferate in a wetter environment, making the growing of some plants even more difficult than previously.

Things might just go on much the same as before, the difference being no one would be wasting time trying to discover something that has already been discovered, famine would cease, houses would remain intact longer, forest fires would get put out safely and efficiently, fewer ships would be lost at sea, people would be happier, etc. So the overall economic impact would probably be quite good, and it wouldn't happen overnight. Some of the economic ramifications of minimizing weather disasters would take years to be noticed.

Businesses might exist after years of peaceful weather that would have stood no chance of getting started without the stability of the weather. Had mankind not taken charge of the situation, the resources that enabled these marginal businesses to succeed would have been entirely taken by the construction industry and the rebuilding of structures destroyed by weather and the building of desalinization plants, and they simply would not have existed at all.

An international committee could be formed as a clearing house for all nation's meteorological plans for each upcoming week, and neighboring countries would be able to prepare for these events as well. The citizens are all told that any unauthorized intervention into the weather by a member of the public without the appropriate permits will not be tolerated. The National Weather Service would continue to give weekly forecasts and give out information on when and if it will rain, and also take active steps toward implementing the new weather modification technology.

Equipment would be required, such as flatbed tractor trailers,copper tubing,some copper wire to tie it down with, some lead, a few helicopters and some manpower to remove outlaw copper or lead placement, and drive and safeguard the flatbed trucks. I'm assuming the meteorological community would already be in possession of a few barometers.

The likelihood of anyone profiting monetarily from this idea is little, other than a rise in the standard of living that would be slow at first but gradually increase, as one thing leads to another, shall we say. Though there are some expenses involved, it is the most economical solution to a plethora of problems concerning the environment around the world. The arrangement of six foot tall copper coils tied together with straight bars of copper tubing described earlier isn't necessary if one has a flatbed truck with a lift feature similar to what a garbage truck uses. Lay the tubing along the bed of the truck in a row tied down securely with copper wire,and park it in a favorable spot. Elevate the bed of the truck to as nearly vertical as possible, facing west. Wait for three days.

Naturally anyone who has an idea they are promoting  thinks it is the most worthy of public funds or should have persons interested enough to invest money in it. In this instance I am no exception. Nevertheless, the problem with this discovery is that no one has any facts or experimental results that one can point to definitively and say- see, there, publication such and such, pages so and so, etc., where some specific persons conducted specific tests with exactly so many pounds of copper and so on. So the truth is really still to be determined, at least as far as I can see.

The overwhelming usefulness of this thing aside, many aspects of which are scattered around various pages of this publication, when I see such a publication and all the attendant observations that were made in each 72 hour period when the experiments were conducted I will be satisfied that the scientific community fulfilled its duty. There is little doubt in my mind that my hypothesis will be ultimately proven correct.

Weather happens all the time. How can one tell if a storm occurred naturally or not? Just because a quantity of copper is on that hillside doesn't mean it wouldn't have rained regardless if the copper were there or not. So we need discerning individuals who conduct experiments with copper when the forecast for rain in the next 5 days is non-existent. When there is a forecast for rain in 2 or 3 days an experiment involving lead might prove instructive. Further research is the objective we have in mind.

If, once the truth is determined, no one sees fit to spend time or money on an idea whose fruition will not be monetary, but rather enhancements to all life on Earth in the long term that each individual will receive in small increments that will largely go unnoticed, such as not worrying about the weather each and every time you step outside, let the market do what it will. I cannot force my value judgments on anyone nor do I intend to. I just hope the science of meteorology means the same thing to me as it does to the other 6 billion plus humans present on Earth. Aristotle even commented on the pleasure of knowing things.

Human beings are under the necessity of sharing the Earth with fellow humans as well as all other living things present on the planet. Assuming we all know that science deals with ascertaining what things are, and how things work, when we find something that works that may prove harmful we record the fact and the governing body steps in to regulate the use of such a discovery.

The Bureau of Alcohol, Tobacco, and Firearms exists to protect the public from the dangers of these things. There aren't just a few people with guns in hiding and everyone else pretending guns don't exist, right? Guns are a device we are forced to live with everyday because someone discovered how to make one at one time and that knowledge, with improvements, is now part of our civilization.

If the ATF became WAFT, weather, alcohol, firearms and tobacco, that would simply be the way things are. So, whether or not it is economically feasible to modify the weather, if we once and for all declare publicly that we are capable of it and it poses a possible danger, one may rest assured that the government will step in and enact legislation to protect the citizens from possible harm.

At present, it appears they have jumped the gun and have prevented curious people from ever finding this out,rather than letting the discovery be made and acting appropriately from that point on, though we really can't conclude that with certainty. I really think Nile Blue or Blue Nile is about something else entirely, and this feature of nature we have been discussing has just been debunked rather superficially and from then on conveniently overlooked, the moral high ground of the scientists preventing them from revealing what they discovered, so there was never really anything with any paperwork involving this discovery anywhere. Nonetheless, all that has been put forth here is simply intended to show that such a feature of nature truly exists, and of course, what that implies.

Marconi stole Tesla's ideas regarding wireless transmission of radio signals, so ruled the Supreme Court, before I was born. It happened that Marconi was one of Tesla's lab assistants during one of Tesla's more prosperous times and that is where Marconi got the ideas he claimed were his own. I rather resent having learned that Marconi invented the wireless in school, almost two decades after the Supreme Court decision. This is just one more instance demonstrating how sluggish governments are when confronted with change. I sincerely hope there are new textbooks now with Tesla credited with the discovery of wireless communication.

We learn something or we don't. Once learned and properly positioned in the science to which it belongs and recorded in Encyclopedia, the task of the scientific community is finished. What happens afterwards with regard to it is up to public opinion and ultimately, court decisions.

I suppose the Third World countries won't be as well equipped to deal with possible threats from abuse of this device. These countries won't have the number of meteorologists we do, nor as many weather monitoring stations, or the advanced equipment such as Doppler radar that the United States is in possession of. Funds may be lacking to expand meteorological activities in some countries. Nevertheless, these countries will undoubtedly benefit as well from this knowledge.

Most of the Third World already has the problem of either not enough precipitation as a general rule, or too much precipitation, or both, at various times of the year. The weather problems they experience is a large part of the reason why a good number of these countries are classified Third World countries in the first place. Bangladesh, for example, would benefit from the even playing field once this knowledge is assimilated and, ultimately, put into Encyclopedia for the curious to learn. Simply the absence of tropical cyclones hitting that low lying country would in and of itself be a great boon to them, as well as their being able to lessen the impact of the yearly dry seasons, which could be easily accomplished without umpteen meteorologists and weather monitoring stations.

It is possible that some countries who were at odds with each other in Africa had positioned lead in large quantities so as to cause neighboring countries to suffer from prolonged drought. Africa in 2008 was in the grips of some of the worst flooding it has seen in a long time. So now maybe the metal of choice has changed between the warring nations. All we can truly say is that the flooding in Africa in early 2008 was one more instance of weather changing to wet abruptly.

That is the entire point of this exercise. Time and time again some anomalous piece of weather information stares us in the face and adds to an ever growing total attributable to which cause, global warming or man made metal deposits? A lot of these instances will never reveal their true cause. A fair proportion of them are probably man made, though. Wisconsin tornados in January 2008, for instance, almost certainly got a little help from some man made device. Nature is certainly capable of any and all weather events or they would not have happened in the first place; the question is would they have taken place without some additional factor entering into the equation?

In light of all that has been discussed, watching any program about nature or the environment on television is quite a distracting thing to experience considering that so many suppositions that are made on the program are negated by the fact that we are fully capable of doing something about it. No full discussion has ever been made of these topics with the assumption before hand that weather modification is possible. Even a simple nature program can offend the enlightened, as for example, a film about certain areas of Africa and the wet and dry seasons and the animal migrations. "And then the rains came." runs the narrative, and at that point one finds oneself wondering what the intelligent beings who lived there were doing in the meantime.

A lot of the global warming documentaries are similarly flawed. A major theme running through most of these are that global warming will change weather patterns drastically, such that wet areas become dry and vice versa, storms will be more severe, etc. That remains to be seen, especially when a means of changing it by human hands exists and has for well over a hundred years. Another recurring theme in discussions of global warming is that we will reach a point of no return one day where runaway global warming will takeover and be unstoppable.

I hope someone takes notes on how many weather events destructive in nature occur after this discovery is placed in the encyclopedia, and compare that total with the totals that occurred before that for a comparable period of time. When a 98% reduction in destructive weather events happens, how wise will keeping it a secret seem? We are at the top of the food chain on Earth. We are it's custodians, we must do the weather, it's as simple as that.

Something large and conductive, in a sufficiently high location, silver, copper, or whatever, will cause a drop in barometric pressure. Clouds will develop. Precipitation will occur after 72 hours give or take 8 hours. Lead or some similar non-conducting material in sufficient quantity in a high location will cause a rise in barometric pressure. Clouds will develop less readily. Precipitation will be infrequent to non-existent. There! We have learned something. Why isn't it in the Encyclopedia? The residents of the Marshall Islands wouldn't have brought drought upon themselves,so why not give these people the chance to continue to live the kind of lifestyle they were accustomed to before other members of the human race drained the oceans of precipitation?

How the government will deal with this issue is important. To say that a legal quagmire looms over this issue would be an understatement. The only way to resolve such problems would be to establish guidelines as to how much and how often precipitation can occur for reasonable amounts of water to be available. It is not too difficult to perceive when an area is experiencing drought. Plants start to shrivel in the dry heat,animals begin to die, and the necessity of precipitation is evident.

Planning ahead and assuring that such conditions do not occur would be a reasonable course of action. When rainfall becomes excessive and flooding starts to occur is also not difficult to establish,so finding the happy middle ground where neither drought nor flooding occur shouldn't be too difficult, nor should actions taken to achieve such goals be subject to litigation by unhappy persons discontent with rainfall totals, if as we expect the authorities assign these tasks to meteorologists, provided no damaging weather occurs.

How many instances of damaging weather will result from initial experiments before the experimenters get it right? This device and how it can be used, and to what quantities safely, and what quantities pose a problem has "proceed with caution" written all over it. Any competent and thorough experimenter would begin with a safe quantity, perhaps three or four hundred pounds, see the cloud development and maybe a little light rain, pack up, return some time later with additional copper, find the storm system to be of the size he is aiming for, and assign the task to a local crew on standby for the weather service.

Eventually the entire country will have standby crews waiting for orders to transport the stockpile of copper, or lead, to the agreed upon site for placement as directed. Before that can happen, though, this theory would need some discussion, and it remains to be seen if humanity can overcome the differences of opinion they all have regarding this issue. One thing is certain, though; ignorance has not been bliss.

There are other implications of this discovery that go beyond the usefulness of the discovery itself. Philosophically, this discovery would not further the goals of those who believe that a God created the universe. Once again, it would point to the sad fact that we as humans are quite on our own when it comes to dealing with things, including the weather, the various manifestations of which are already classified as "Acts of God" by the authorities.

The intention of this book was never to champion atheism. But,as the Boy Scout hopefuls pointed out so aptly who were excluded from joining the Boy Scouts because they were atheists, we do not have His signature on anything. It has been noted that there has never been a President who did not side with the notion of a creator. So the mainstream thinking is not going to find this discovery useful in forwarding their agenda.

My problem with that is truth being lost in the shuffle. In the true sense of the word, the notion of a creator is still an unproven assertion. Beside that the whole notion of faith is flawed. The famous quote goes, "To Doubt Is To Think". We as a species are not faithful beings. Human being are curious beings by nature, and discerning what is true or false is important. We are the only creatures known that are capable of thinking. The very definition of man is "thinking animal". So to have faith in something that one has no proof of is contrary to our very nature, and a direct insult to a God, if there is such a being.

The path to living in harmony with God, if there is such a being, would be to exercise those talents with which we are born and that come naturally. Then, upon one's eventual demise, if we are really confronted by the Pearly Gates, who do you suppose would be allowed to enter, those who used their minds the way they were supposed to be used, or those who simply followed instructions? So, in this day and age, the Inquisition is still operative and keeping enormously useful gadgets from coming to light.

Leaving no stone unturned in the search for God turns up the most enormous universe perhaps, an unknown number of times bigger than our known universe, looking and acting like a wilderness in every respect, if we assume that what is beyond here is more of the same as we have here, with some really big black holes scattered around that we probably don't have here because two collided in this tiny quadrant of the entire universe to create our known universe and the nearest one now may be near some of the largest super clusters in our known universe.

No sign of God in a known universe that contains an estimated 100 billion galaxies. Our Milky Way has an estimated 100 billion stars. Considering that it would take quite a bit to create 100 billion galaxies and still more to create unknown multiples of that, God would certainly grasp our rules of evidence and respect them, these rules of evidence having been invented by our minds, which he created. Knowing that we are curious,able to doubt,and have limitations,he would have certainly signed a document for us, maybe even have explained in the King's English where he has been.

Probability of actually proving God's existence much lower than winning Powerball. I'm only presenting this evidence for the sake of completeness. Certainly anyone who decides to believe in whatever God is free to do so. I've never seen each of the 100 billion galaxies in the universe, but in this day and age, if an astronomer were to discover God near or in one of them, it would hit the news, what?

A knee jerk decision made by some 1910 politician is going to cost a lot of money over the next few decades unless some truly spirited effort is made to revolutionize the handling of the weather. It can be done, it has expenses, but the savings it provides, the increased availability of water that would come about along with all the other benefits discussed previously make the choice clear.

I don't really know if this feature of nature was ever noticed by anyone besides Tesla himself. Maybe there has never been a deliberate concealment. That would be in keeping with the long time it took mankind to recognize electricity in the first place. Nevertheless, it has been at least discussed by some meteorologists, as evidenced by the one paragraph from our 1980 book with the title Weather Modification. So it could be that an honest mistake was made, previous to 1980, and now with the current theories abounding about dark matter and dark energy it is an idea that is screaming for a closer look.

Let's return to one of the first sentences of this book, the stance of the meteorological community toward weather modification. Actually, there are three to five distinct methods that we know of that can change the weather. Electromagnetic pulse, a plume of conductive compounds sent airborne from a furnace, chem-trails, and strategically placed nearly pure conductive or non- conductive metals of some size. We could have the Chinese for a fifth, with their attempts to stop rain by firing bombs with water absorbing chemicals at clouds during the Olympics, though whether or not what they did was effective is unknown to me.

Three to five distinct weather modifying methods conflicts considerably with the notion that only cloud seeding is known to help a little to generate precipitation. There are many ways to skin a cat! As to which method works best and is most cost effective, the metal placements in strategic locations wins. I suppose there are probably a lot of meteorologists who would agree with me after reviewing what has been stated here, so the official stance isn't necessarily always endorsed by everyone.

Knowledge for knowledge's sake is an important part of living for a lot of people. Many non-professional people have a keen interest in science. They specialize in some other area in our division of labor society. They expect the findings of the professionals occupied with the sciences to be eventually reported to them. A shipment doesn't just not arrive from somewhere. Why shouldn't what has been discovered be reported to them? After all, 110 years is a long time, from the first copper meets atmosphere events, more than enough for any patent to have expired long ago.

Scientists need to fulfill their end of the bargain, and faithfully report in greater detail the outcome of experiments such as we've described. Maybe fear of widespread panic by the general public keeps this knowledge from public discussion. That doesn't help the fact that any number of coastal homes could be underwater in a few decades. I don't think the public would panic. If it turns out we can modify the weather in more ways than one, so be it. We integrate the new information, then move on to the next unknown.

A thorough examination of what might befall the world should this discovery be announced to the general public should consider just what the general public is capable of. A profound discovery such as this would generate a lot of interest. Does that mean that x number of people with pick-up trucks are going to try this for themselves? I would guess it means that more people would buy barometers and watch the weather channel more often. It also means that any unusual fall or rise in barometric pressure that has not been planned by the meteorological community will be instantly reported by hundreds of home bodies, perhaps thousands, people who for whatever reason spend a lot of time at home idle, or doing housework, and able to watch television.

The acceptance of our mastery over the weather and the excellent weather that we all begin to experience will eventually sink in, and after a few years, the number of people watching the weather channel because of the novelty of that new discovery will begin to tail off, and probably continue to tail off so dramatically that eventually it could disappear completely, with regular news channels giving tentative weather plans.

The world wide meteorological community would all see that world wide participation would be most conducive to peaceful weather, since decentralizing moisture from the oceans by using numerous copper placements around the world would prevent huge storms from developing. Who will take care of the polar regions will ultimately be decided, and plain ordinary rain could get more frequent. If that is all that happens and everyone knows when it is going to happen, it becomes commonplace; some young people may opt to enter the meteorological field, most people will completely lose interest once singular or calamitous weather events hardly ever happen.

Tesla arrived in Colorado Springs on May 18 of 1899, so give it a month for preparation and sometime in June,1899, we can break a bottle of wine and commemorate the event, the first known insight into this feature of nature thus far described herein was glimpsed. The brief mention in Tesla's Autobiography of Colorado Springs that I saw was hardly complete, at least from the meteorological perspective. Certainly there are historians who have studied Tesla, there is still a Tesla club of some kind. Perhaps if enough people learn of this, we will get a more detailed account of what transpired in Colorado Springs.

The U.S. military, as before mentioned, has a weather modification project classified since 1945, titled Blue Nile. What's up with that? Is it still classified, and if so, would it be so untoward of the civilian population to ask that it be declassified, in the hopes that information therein might help resolve problems with the environment,which are a priority now? Water shortages on a planet that is 70% covered by water. In a universe where every last primordial speck from the Big Bang hasn't condensed into hydrogen yet. Where the creature living at the top of the food chain has resolved the problem of how to distribute fresh water cheaply and efficiently, yet cannot bring the matter off, as it were.

The longer the human race delays in recognizing the feature of nature that is there to be used by those who truly face a shortage of water,the worse things will get. Things could get a whole lot better. The grass could be greener on every side.

As time goes on, and this type of weather modification activity increases, which may be happening already,it will become increasingly more imperative that all areas of the globe be aware that they may experience a shortage of water. In spring and summer of 2009 there was a drought in Northern China; a drought in Australia with severe brush fires; various provinces in India did not get a monsoon this year; the United States has two or three droughts according to the Weather Channel; Brazil, of all places, is also experiencing drought and this type of situation can only worsen, if everyone remains ignorant but a few.

Naturally not everyone is going to agree as to how much precipitation should be occurring in the area in which they live. The air and water are public things; access to natural bodies of water is usually made a part of law, no one can prevent another person from having access to a lake even if they own all the land surrounding the lake. Lakes, oceans,and rivers are not bought and sold; they are part of the public domain by law, just as sidewalks are. No one can buy part of the air and prevent others from breathing it, either.

How is the issue resolved as to how much precipitation should occur in a given area? People who suffer from arthritis hate it when the barometric pressure falls, they experience more pain then, all the joints in their bodies ache, they can barely open a screw on cap because of the pain in their wrists, walking can be torture for arthritic ankles during a period of lower barometric pressure. People who ride motorcycles, too, never seem to want any rain to put a damper on their enjoyment of motorcycle riding. Others need to use a motorcycle or scooter to commute to and from work, and a rainstorm is simply not acceptable to them.

Water is a necessity of all living things, and some precipitation must occur if life is to continue at all. So it looks as though some people are going to be inconvenienced from time to time to facilitate the continuance of living things on land. Having accurate advanced forecasts would help immensely. At least those who are the most inconvenienced will be prepared for rain on schedule.

Economically, the allocation of scarce resources changes continually as circumstances change. Rebuilding structures that have been destroyed by weather may provide jobs, but as far as improving the productivity of labor and increasing the wealth of the citizens,it would appear that retaining those structures that are already built and hence permitting the allocation of resources to new developments would accomplish much more.

A dollar figure would be hard to arrive at when one considers how many structures have been destroyed by weather or forest fires over the past century, and the labor and materials needed for rebuilding these have been withdrawn from other areas. A long period of time with little severe weather anywhere on the globe could usher in an age of prosperity never before seen, along with bumper crops year after year.

If it is possible for human beings to take charge of some aspects of nature and control what it does, do you suppose that will ever happen? Volcanoes will still erupt, earthquakes will still happen. The possibility that the Earth could be hit by a large asteroid one day is still real, but horrendous weather could have disappeared about 80 years ago, and should certainly be gone now.

Summarizing, Tesla in May or June of 1899 mounts his huge Tesla Coil on the roof of his new laboratory in Colorado Springs, Colorado, Hatfield shortly after seeming to be successful in his attempts to make rain by sending plumes of copper compounds airborne. Starting then the United States goes through three decades of wet weather followed by almost a decade of dry weather. The U.S. Military classifies Blue Nile, or Nile Blue, in 1945. Ayn Rand's "Atlas Shrugged", about which we'll discuss more in chapter seven, is published in 1957. My experiment with ambient static electricity brings me to an awareness of the possibility of modifying the weather in 1980, various droughts occur, I write letters, flooding happens shortly afterward.

Once I got a computer, I e-mailed a lot of meteorology students and professors when I found e-mail addresses on College and University Meteorology Department web sites. The east coast has a lot of universities. I wonder if that has anything to do with the rather cool, wet summer there, and increased precipitation over the past 10 years or so. No reply from meteorologists but one, unless you include "Who are you and why are you sending me this stuff?", until recently, now that this writing has progressed some. Before 2007 my description of weather modification was too brief and mostly ignored. So this book is the only way, really, to get my point across. No science journal will publish non-scientists work, or only very rarely.

Actually, if this presentation is proven correct in the end, one could argue that it should be required reading for first year high school students. It might teach the youth of the world more than just weather modification or cosmology. They may learn and adapt, and future civilization might no longer see the lethargy and apathy towards issues affecting the environment, government and the general public, which the last three or four generations of human civilization have shown. Would that make us ultimately more like honey bees if we were better able to do things in concert? Or a colony of ants? I think it would show we grew smarter.

Chapter 5. Benefits of the New, Costs of the Old

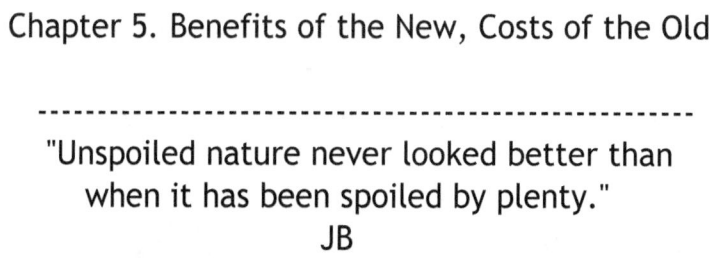

-----------------------------------------------------------

"Unspoiled nature never looked better than
when it has been spoiled by plenty."
JB

-----------------------------------------------------------

Desalinization plants have been built, are in the
planning stages, or are being proposed in a number
of locations around the world. The expense of
building a desalinization plant for fresh water far
exceeds the cost of timely placement of copper
near a mountain top. Desalinization plants also will
do nothing to mitigate flooding, which could also be
readily accomplished with this undocumented
discovery we have been discussing. The polar ice
caps are melting at an increasingly faster rate, and
the only way to counter this potentially devastating
inevitability while reducing fossil fuel emissions
would be to load the poles with copper during the
winter time at the respective poles and hence
cause more snow to fall thereby rebuilding the ice
caps.

The Earth as an ecosystem has a total quantity of
water vapor and static electricity. Diverting a huge
quantity toward the poles while additionally

diverting more water vapor and static electricity toward land, based on meteorological data and rainfall totals, would leave the oceans without the necessary accoutrements to develop hurricanes, cyclones, typhoons, etc. The beast will have been tamed.

The increased runoff from land would be beneficial to the marine organisms on the planet. I saw a nature program once where the program pointed out that the source of almost all the iron in the sea comes from precipitation runoff from land. Coastal areas are important to the overall health of all marine organisms. The larger creatures that cross entire oceans are dependent upon coastal prey for iron intake, which is probably as essential to marine creatures as it is to land animals, or nearly so. Rivers like the Colorado, that currently are almost completely used up by the time the water reaches the mouth of the river, are not helping to improve fishing worldwide, or the health of the creatures living in the oceans.

The tiny trickle of water that reaches the Sea of Cortez at the mouth of the Colorado doesn't do much to enrich that coastal area with iron. My hope is that once this idea becomes accepted as fact and made proper use of, the trickle coming from the Colorado would eventually be a more robust runoff, and coupled with other means of generating electricity besides building dams could eventually make possible the tearing down of some of the dams already built, returning some river systems to their natural state. Perhaps as dams wear out, and keeping the dam versus adopting new technologies is weighed carefully, one by one the dams will come down.

Plants and animals in the wild, some near extinction,would all begin to have an easier time if water were more wisely distributed around the world. When one looks at what mankind has done to feed the growing populations of humans over the centuries since the Americas were discovered, it seems the least we could do to help creatures in the wild. The North American Bison nearly went extinct. The dodo bird and passenger pigeon are extinct along with a number of less well known creatures and plants. Amphibians are thought to have suffered over 100 extinctions over the past century.

The entire downward spiral of living things in the wild that has been going on since mankind began encroaching upon territory that was once occupied only by wild life could be reversed. It surely adds to one's enjoyment of nature if one actually sees wild animals and plants in great profusion if one were to travel to a national park. The fact that humans happen to be micro managing the weather to provide water for the living things there isn't going to detract from the beauty of the place, in fact, it would help a great deal in helping wild things grow, and one's chances of seeing some creature or plant that is rarely seen go up.

So, no trace of what is taking place with the weather would appear as an unwelcome change in the wild. The modification of the weather involves no building of fences or any other thing to impede the movement of animals in the wild. The fact that something is placed somewhere in the area periodically poses no danger to the living things near it.

If, for example, the desert southwest of the United States embarked upon a weather modification program designed to provide a decent sized storm about every ten days during spring and summer, and perhaps once a week in fall and winter, the populations of desert rabbits and tortoises, various birds, and almost all the plants in the area would begin to increase. After a few decades it might even be possible to hunt rabbits in the southwest. Mountain lions would also see an increase in populations with fast reproducing animals like desert rabbits in greater abundance.

After a while, driving through the area could prove a lot more interesting with more plants and animals around. Eventually,problems will inevitably arise when some wildlife encroach upon human settlements, but that is something that happens already. Moose walk through Alaskan towns routinely, and deer, bear and lions occasionally wander into towns all over the United States. Some of the countries where monkeys live have the problem of dealing with thieving monkeys everyday.

So incursions into human areas by wildlife could increase with increasing populations of wild animals, but these incursions might possibly decrease because the animals are finding enough to eat in the wild. Bears won't need to raid garbage cans if other food is available in the wild. Bears don't visit human habitations for company. They are there to eat, and mostly because the food in the wild is insufficient at the moment.

The whole question of what becomes of wilderness areas and how weather modification activities will impact upon populations of wild animals and plants will just have to wait until enough time has gone by to make some kind of assessment. With the use of such devices as described here, there is little doubt but that increased populations of living things in the wild will result.

As to whether this increase in populations causes further problems for humankind, that is entirely possible. Perhaps we will see more locust plagues, more problems with nesting birds, more outbreaks of mosquito -borne illnesses, etc. Inhabitants adapting to locust plagues could catch and deep fry the locusts. Tasty. None of the possible problems that wild populations could present would be problems humans have never seen before. But then, wild populations have never had weather conditions optimized for them before.

Perhaps some creature will become way too numerous and pose such a problem that the only way to stop this problem creature would be to stop any and all precipitation in a geographical area until the creature dies off. Fortunately, the theory gives us a way to do that. Any runaway population explosion of some disease vector animal or whatever kind of problem plant or animal will at least  start out locally, and if local health authorities are vigilant, could be stopped before spreading around the world.

The whole idea of interfering with nature will meet with opposition, though only benefits could be seen as a result. Even global warming fanatics will not welcome this theory even though it introduces an easy solution to that problem. Nature, however it comes, is for some mystical reason better than having perfect growing seasons, an absence of damaging weather, and higher polar ice caps. The fact that optimizing the weather would cause populations in the wild to double and triple still is no substitute for the real thing, that which happens if man does nothing.

The increased use of this device would be a great help in reducing air pollution. Major cities with smog problems like Mexico City and Los Angeles could plan on more frequent showers to cleanse the air more often. Most of the carbon dioxide would end up in the oceans eventually, but at least the atmosphere gets cleansed of it and other air borne particles. People with asthma would have an easier time.

Precipitation is about the only way air pollution can be reduced, once the pollutants are airborne. There has been some research on funneling air into $CO_2$ scrubbers that remove the $CO_2$ that comes in and release air free of it, but whether that idea will be developed on a large scale remains to be seen. But at least we are beginning to see some scientists looking at things that could be done on a large scale, encompassing planet wide activities.

We may see the day when several dozen huge $CO_2$ scrubbers are put into use around the world, along with a more honest weather modification approach by mainstream meteorologists. Every additional rain storm, or even cloud over land slows the global warming problems,postponing doomsday a little longer, until perhaps we will have got ourselves out of the mess.

Weather forecasting could become a planning event rather than a guessing event. Should a rain event be planned, there would be no reason not to tell the public about it; hence people could plan accordingly. Boating outings could be planned for the days between rain events. This could be a revolution in advanced planning.

After twenty years or so, organizations like professional baseball might decide to schedule games and days off in sync with the weather forecast, which by that time would probably be 100% accurate. Land heats up far faster than water during the daytime, so further extended use of this gadget would tend to increase cloud cover over much more land area of the planet, cooling our planet, and also slowing the threat of global warming.

Forest fires have been fought in some instances for weeks, with fatalities occurring among the fire fighters. The National Forest Service continually monitors all forested areas, and are usually aware of a forest fire within an hour of its origination. Given the wind conditions and the likelihood of the fire becoming difficult to contain, copper could be in place beginning its seventy-two hour storm incubation within three or four hours of a possibly dangerous fire.

Actually with use of this by properly cautious meteorologists few areas would be prone to forest fires because most places would be receiving enough precipitation to keep the vegetation moist, and unlikely to burn out of control, although in the case of arson or accident a fire would still be possible between planned rain events. Windmill farms could use horizontally placed copper to augment the wind and increase the yield of the windmills.

Shipping could see a marked improvement worldwide, since so few ships would be lost at sea since the shipping lanes could be freed of storms. A ship laden with a ton or two of lead on its deck could patrol the edge of shipping lanes and intervene between approaching storms, causing them to either change direction or weaken in intensity, or both.

The ocean is basically flat, and barometric pressure is on average much lower out over the oceans than on land, where the uneven terrain, the lower percentage of water evaporating skyward and the more extreme temperature changes between night and day all contribute to rising barometric pressure.

Diverting the miniscule, free floating molecules in our atmosphere toward land is possible, and together with the diversion of water vapor towards the poles would cripple the large storms we see now, and make ocean travel safer in and of itself, even without a vessel carrying lead on patrol.

A look at the Sahara Desert and the Amazon Rain Forest reveal some interesting metallic differences between the two areas. The Amazon sits just east of the Andes Mountains, where 70% of the world's copper ore can be found. Copper ore is only around one percent copper, in contrast to copper refrigerator tubing, which is more than 99% copper. A number of bits of evidence point to the Sahara Desert having more than the usual amounts of lead at or near the surface of the land.

The Ancient Egyptians used cosmetic face makeup countless centuries ago, mostly consisting of lead based paints. Mummies unearthed have been examined and have been found to possess high concentrations of lead in the bones. It has even been speculated that a fair proportion of these people died of lead poisoning, or,at least had their lives shortened by excessive lead accumulating in their bodies.

Drinking from earthenware vessels that were made of clay with excessive amounts of lead is thought to be the cause of the accumulations in the bones of these mummies, since the lead would leach out of the clay into the liquids being drank, and hence into the bodies of the persons drinking the liquids. The reason the lead is more abundant there is because that land area has some of the Earth's oldest crust, that solidified originally 4.5 billion years ago, and has never been torn asunder by volcanoes.

Between the Amazon and the Sahara we find the Sargasso Sea, the place where ancient mariners were becalmed for weeks on end, and since wind is what the vessels they were sailing on depended, this was a most difficult part of the ocean for the ancient mariners to traverse.

A meteorology professor replied to an e-mail I sent to him that I obviously didn't know anything about the science of meteorology in general or the weather patterns of the Southern Atlantic in particular, and that everyone knew that the notion I was promoting was a croc. I didn't put the rain forest next to the copper (the World's largest Rain Forest) or the largest Desert in the world on top of an unusually large amount of lead.

The laws of physics dictates what occurs, if the evidence supports it, it must be true. That is the difficulty with weather phenomenon, at present, since this particular discovery could be easily overlooked and attributed to mere coincidence, starting with Tesla and his experiments in 1899. There have been far too many "coincidences" already to justify rejecting the particular concept that the weather can be easily controlled with a little time, patience and perseverance, to say nothing of having the truth of the matter on record.

Either that particular meteorology professor is blind, or doesn't want the truth revealed. The ivory tower stance that is so prevalent in the scientific community seems to be operative here; the possibility of abuse of this feature of nature by just about anyone justifies keeping over 6 billion people in ignorance.

The means to build an atomic bomb,for example, is not easy to find. Manuals describing how to do this are not available to the general public. The difference between building a thermonuclear device and simply putting some copper in a high location is quite striking; one is a very complicated and difficult undertaking requiring some rather rare and restricted items, along with the means to assemble them properly, and the other is remarkably simple. It is this simplicity and the likelihood that abuse could be prevented that indicate it would be wiser to permit this discovery to become available to the general public.

A nuclear bomb can only be used for destruction or the prevention of war. The awful effects of nuclear war has prevented all out wars from being fought. The discovery of how to control the weather has countless peaceful purposes. World population tops 6 billion, global warming becomes an ever growing problem and the number of people trying to discover a means to counter rising temperatures, melting ice caps and rising ocean levels among other environmental problems is now in the thousands,perhaps even a million or two. Or should we just tell these million or two people to stop trying to solve these problems since whatever solution they come up with won't be used anyway, since whatever it is, if it works, it will work in excess as well, and be unacceptable in the ivory tower?

Rising Ocean levels is the prediction of the global warming faction,those who think global warming is happening, as opposed to those who think it is not. All relevant information has not been accumulated, or at least not enough to know with certainty what the future holds in this regard. There could possibly be nothing to fear; our puny efforts are overwhelmed by the inexorable changes that the Earth goes through, and temperatures could start declining on their own.

More likely, humanity will do nothing and temperatures will continue to rise. How long before everyone realizes it is too late to avoid some major problems with coastal cities is unknown. The chance to actually avert such catastrophes exists, and should be utilized, even if temperatures may come down eventually without human intervention.

Global warming is a good thing for people in cold climates, and if it eventually happens,if it happens slowly enough the end result before some ice age or another would probably be good.

If the Earth is on a warming trend,with or without humans burning fossil fuels, there is little we will be able to do about it but reap the benefits of longer growing seasons and more arable land. Efforts to maximize polar ice caps and add more water to the land will slow the warming trend down some,perhaps enough to make the transition easier.

We can also throw in the cost of continued searching for dark matter and dark energy, when computing the costs of letting this idea slip through our fingers, if it ultimately is proven that more hydrogen is coming into existence near the path of least resistance created by the existence of a strategically placed row, or rows, of copper tubing.

One experiment to confirm or deny this could consist of: an experiment with copper such as has been described, with random air samples taken after 72 hours, followed a week after removal of the copper with an experiment with thin lead sheets, all placed the long way north-south so the entire face confronts the prevailing westerlies,with air samples taken after 72 hours also. A determination of how many hydrogen atoms from each series of samples of the same size could be made, and if the copper samples showed a higher concentration of hydrogen,the missing dark matter and energy are found.

Well,we wouldn't have exactly found the missing particle, but we would know that it would accumulate in greater concentrations along a path of least resistance and that it is there combining with others of it's kind to become hydrogen. That may be the only way we can confirm the existence of a primordial speck, if it eludes detection as itself. They do rather neatly fit into an explanation for dark matter and dark energy. Newer equipment designed to capture a hydrogen atom appearing as though from nowhere might eventually be created. However many millions and scientists, buildings, equipment, etc., are expended in the search for dark matter and dark energy could be expended on more promising pursuits, though there is the likelihood that most research will continue anyway,since the problem is still unsolved.

We also have people searching for a solution to these problems with the weather that we are contending here have already been solved. How much time, energy, capital and equipment is being thrown into such pursuits is probably considerable. The burden of proof to me falls to the meteorology community, or at least to the astrophysical community to ascertain the results of the experiment just described, whether hydrogen is more abundant in the copper part of the experiment and less abundant in the lead part of the experiment. It behooves them to conduct such an experiment.

The costs of living with nature as it comes will rise, there will also be the stunted intellectual growth of children to consider. Concealing pertinent truths from children isn't wise. The more we are free to learn, the more we can prosper thereby. Besides the worsening forecasts for upcoming decades, there is still the truth, and if it isn't addressed soon, anyone who cares to try a strategic placement of copper or lead of their own won't be held accountable for anything in the event that they are seen doing this experiment. They will be forced off public land. Probably that would be all.

Mischief such as described might not all be mischief, if no one else seems to be trying to do this, why don't I? So everybody and his brother would think. We need rain. That kind of thing, if left unchecked, could prove disastrous if too many people got involved. So I think the grim truth must soon be faced for the world to see better times rather than worse. The only way it can succeed is by more and more people learning of it.

Eventually it will break above ground and sprout

 beautifully, and take its place amongst other great discoveries. It will become an accepted part of the routine of living. Nearly everyone will be on the lookout for any unplanned changes in barometric pressure. The absence of severe storms and an easy way to distribute water to every land mass on the planet would make it possible to prosper as never before.

## Chapter 6. Looking Ahead

------------------------------------

" The happy middle ground
is more fertile"
JB

------------------------------------

I think that any country that begins a weather modification program such as this gadget we have been discussing should conduct sweeps in a helicopter with a barometer in any area of that country that has seen a lot or very little rainfall in the past decade or so for man made metal deposits. Nuclear power plants with thick lead walls to contain radioactive materials in the event of an accident are a factor to consider since the presence of that much lead will tend to raise the air pressure.

Copper or lead may have been placed in remote locations and abandoned, and these lead or copper placements would need to be found. Indeed, the desert southwest of the United States has remained unusually dry for quite a while, and the last nuclear plant built in the U.S. was in the San Diego area.

Arizona, Nevada, and Southern California are the favorite retirement locales of a number of people. It would not surprise me to find that the desert southwest may just be one place where lead has been positioned and left there by some arthritic old codger,who is probably long dead now. This person may or may not have had a military background, it may have just been someone who, like myself, put two and two together.

For instance, the heaviest rainfall total recorded in the U.S. in a 24 hour period happened in Holt, Missouri back in 1947. I happened to drive by there a few years ago, and noticed a railroad track running through the area. Suppose a huge shipment of copper on open flat bed rail cars happened to be sitting on a sidetrack there for a few days,and then the intense deluge that happened came along. Could it be that someone noticed that? I realize that this is only speculation, but if this feature of nature really exists there is no preventing someone from making the causal connection when they witness copper in large quantities and a storm that followed three days after.

The earliest smelters of metals could have even picked up on this back in the bronze age. It would be interesting to see if some translations of ancient writings may have been interpreted wrong and actually refer to precipitation and metals. There was extensive irrigation in Roman and Mayan cities, could it be they knew they could depend on rain happening?

What happens to the cacti? Cacti do not find water inimical. They will develop a means of employing the excess water. Cacti will evolve. We will be able to witness evolution. Some of us will preserve specimens in greenhouses. We will be able to witness the differences between the two evolving species as they happen, the one in the dry greenhouse, and the other out on the now lush Earth.

Actually, the possibility that cacti could go extinct is real, with the exception of those in controlled environments such as a greenhouse. They could be overwhelmed by invading plant species that grow much faster to the extent they could be choked out of existence. Desert locations might want to proceed slowly and gradually with increased precipitation to allow local fauna to acclimatize itself to the change, and to slow invasive species.

Suppose one country wants rain and a neighboring country doesn't. An international committee could be vested with the power to decide the proper steps to be taken based on average rainfall totals and the committee should strive to see that such countries that have always been chronically short of water receive a generous additional provision to try to improve the availability of water in areas that are usually too dry.

I'm sure that once the responsibility sinks in around the meteorological community, the proper steps will be taken ensuring that rainfall totals of the storms created almost always fall within the one-half to one inch range, and that is all that usually happens, with the exception of some thunder and lightning. It is difficult to ascertain how often intervention would be necessary; once a chronic shortage begins to occur it would certainly behoove the peoples of that area, or the meteorologists entrusted with that responsibility to ensure that crops do not shrivel and die, wells run dry, etc.

If such shortages do not occur, well, so be it. Action in the form of generating storms would not be necessary. How many firemen sit around firehouses doing nothing because there are no fires to put out? They are a safety blanket, they are there in the event of an emergency. The time has come to be prepared for weather emergencies in a similar fashion.

It would not be an emergency requiring people on immediate standby, suited up and ready to go like firefighters, but rather a group of people ready and willing to suspend what they are ordinarily doing for three or four days, place some copper tubing or lead sheets in an appropriate location, wait to see the results,and eventually dismantle the copper,or lead, as the case may be, and return to their usual lives. In the event of a forest fire, where the heat from the fire and all the dust and smoke may interfere with cloud development, it may just be necessary to position more copper than is usually necessary to produce some rain.

The extraordinary minuteness of the as much as 96% of the universe that is dark matter and dark energy, and the astounding abundance of these likely charged particles means that they are much easier to influence than the stable isotopes of nitrogen and oxygen and the other 1% of the known atmosphere and this raises the possibility that soothing music sufficiently loud in a high place might be feasible also, with the sound waves creating a smooth path, a path of least resistance.

Then one couldn't discount the rain dances of North American Indian tribes, or the possibility that human brain waves could be enough to begin a cascade of particles sufficient to bring about lower barometric pressure. If human brain waves could do this then there could be instances where copper or lead wouldn't do what we expect it would, due to enough human brain waves countermanding it. All we need do is think the weather we want, and it happens? It would sure be cheaper than moving heavy amounts of metals to high places.

Maybe placing copper somewhere just once on high for three days and informing the public so that local people witness the placement and remember the rain that followed and have vivid memories of the copper sitting on that particular hillside, the memory of enough people recalling it once a week maybe could alleviate any water shortages over for the region for a decade. A video of the event could be replayed in a public outdoor theatre once a week.

Once the paradigm shift is made,there will be no turning back. The encyclopedia must convey the truth of what has been discovered in the field of meteorology. Once that has happened, everywhere people will see meteorology in a whole new light. Slowly, the entire economy of the world will begin to grow in a stable world where the flow of goods and services is augmented countless ways by this new discovery.

For example, a village in an impoverished nation sees plentiful rain for a few years. A family decides to open a restaurant in the village, something that has never been done before because of the dry seasons that normally occur. No water, no way to wash dishes in a restaurant. It makes possible any number of things, the water to fill the radiator in a car or truck, one crucial for trade. One could go on indefinitely with examples, especially where the existence of a chance to grow things would lead to further possibilities.

The number of well intentioned people will always outnumber those with other intentions. When the entire world is watching just about every weather event the likelihood that all would go well and according to plan is high. Unfortunately, preparedness for misadventures with this type of thing is non-existent. Raising awareness about what may really be taking place when disastrous weather strikes needs to begin, and facing facts head on rather than conveniently missing them would help.

Scientific verification lacking, and myself being unable to provide anything more than anecdotal testimony, this feature of nature that has been described here may just continue to be completely ignored. I hope this is not the case. We've reviewed the possible causes, the costs and risks, as associated with doing nothing. Now it is up to others beside myself to look over this possibility, hopefully as many people as possible. Since the whole idea of modifying the weather involves an incursion into public things, the public should certainly be the first to be aware of it.

What could go wrong with the environment if the Earth continues to heat up? Rising ocean levels will put many coastal cities underwater, and render them uninhabitable. The cost of relocating millions of people should be added in with the cost of desalinization plants when one compares the cost effectiveness of using strategically placed metal deposits versus ignoring this feature of nature we think exists, and continuing as we have.

Really needed is more clouds over land. The easiest way to accomplish this is to inform the public of what has been discussed here,and let the matter enter the public forum. No doubt a few experiments might be attempted, in time the authorities will respond with legislation. A recent study suggests that rising carbon dioxide levels in the atmosphere which are absorbed by the oceans in vast quantities may make the oceans too acidic and cause, perhaps, mass extinction of a sizeable percentage of marine life. A lot of the coral reefs are already struggling, and this would worsen.

Nothing written here will help any to reduce $CO_2$ emissions,other than to remove it from the air with rainfall. Reducing $CO_2$ emissions will take some time yet,and hopefully some new insights into how to do that. Coastal cities by the thousands relocated, buildings, people, and belongings would ring up a pretty large price tag, as we mentioned earlier, so the construction industry will again absorb a lot of capital, eliminating marginal businesses left and right. Luxury items once possible are no more.

Price figures on desalinization plants are somewhere around 40 or 50 billion dollars apiece. Add the occasional forest fire that gets out of hand, drought,flooding,hurricanes,tornados,all stronger than ever, and we're broke! The world economy cannot afford not to have this simple solution working. Perhaps the meteorologists simply missed this thing. For all these  years I expect the vast majority of the profession never gave it a second thought. There was that one paragraph debunking the whole notion in that one book, that is all that occurs to them.

So we really can't blame Joe the weather guy for not being aware of this. Even the U.S. government probably now consists of maybe one or two people who ever knew of Blue Nile, or Nile Blue as I've seen it referred to, the weather modification program that began in WWII, and was still classified in 1980. No one is trying to blame anyone, what matters here is that the world has changed considerably over the past century, and a discovery of this magnitude must be explored in its entirety.

## Chapter 7. Anecdotal Evidence

---

"If seeing is believing,
knowing is perceiving"
JB

---

A Russian born lady changed her name to Ayn Rand, developed an entire philosophy almost from scratch by herself that was based upon reality and reason, and wrote a number or books, both fiction and non-fiction. Her fiction books had philosophical overtones. Do you suppose it possible that she was trying to point the way as a philosopher because she had some insight into the workings of things that other humans lacked? This discovery we have been examining suggests itself, though Ayn Rand never wrote anything about it, as far as I know. Her best known work was Atlas Shrugged.[7] In that fiction novel there is a discovery, it involves copper and the atmosphere, but there the similarity ends. Or does it?

For Ayn Rand to write a suitable mystery novel with philosophical overtones, poetic license took over, and the details of the discovery in the novel were left out. In the novel, instead of placing copper somewhere on high, and causing precipitation while at the same time capturing lightning and converting it into electrical current, the inventor had discovered a way to capture and convert the ambient static electricity in the atmosphere into electrical current. It is a long book, with a lot of character development.

It is in the character portrayals that most of the philosophical overtones shone through. The inventor, John Galt, is mysteriously missing, and so is this purported discovery. When the heroine Dagny finally catches up with him and the two are together in his hideout in the mountains of Colorado where the invention provides the electricity for the small village there, there is even then no elaboration as to how the thing works. So without going into too much detail, this novel can be construed as a kind of anecdotal evidence. If you haven't read Atlas Shrugged, do so.

Further indications that Ayn Rand, in writing Atlas Shrugged, gave subtle indications about this idea we have been entertaining here, are as follows: Dagny tries to track down John Galt, finds he worked at the Twentieth Century Motor Company somewhere in Wisconsin, visits the now abandoned factory, finds some notes and some copper tubing. When Dagny follows another airplane into the mountains of Colorado, the other airplane carrying the scientist she had hired to solve the problem of how to convert static electricity into electrical current, she loses the other plane in the clouds, and unable to see land, comes to a rough landing precisely in the village where this invention is operative. She visits someone in New York,too late to stop him from abandoning his current life and joining John Galt in the hideout in Colorado. John Galt had just left the person Dagny visited. At this time,hard rain was falling in New York. Galt's best friend Francisco was enormously wealthy through family ownership in copper mines in Chile. There is even a part where Francisco deliberately mismanaged the mines and the stock market crash it brought on was played out. So there are clouds, rain, copper markets,and copper tubing in the web of intrigue in the book.

Though absolutely certain that the results that I claim will indeed occur, as for proof, that is less easy to provide. In 1980, when I happened to first perceive the effects that I have been describing, a major drought was ongoing in the Southern U. S., and I wrote a letter to the Society of Separationists in Austin Tx, explaining what steps could be taken to bring about a storm front. About a week later I saw headlines about heavy storms in Texas and adjoining states with Austin being among the hardest hit places.

These events just described really happened. However, they do not constitute proof that someone in Austin acted upon my suggestion and acquired a quantity of copper and placed it on a hillside facing the Gulf of Mexico, where the warm water generates copious quantities of water vapor. My opinion is someone did act on my suggestion and positioned far more copper than was really necessary to bring an end to the drought.

Or, being so close to the Gulf of Mexico is one place where only three or four hundred pounds of copper would be sufficient. The very thing humanity should by now be able to avoid appears to have occurred. This was the first letter I wrote to anyone, and of course there was no reply. The possibility that flooding might happen had never occurred to me, and the whole news coverage of it left me stunned.

The Society of Separationists was an organization devoted to the separation of Church and State, started by Madelline Murray O'Hare, who fought some famous court trials concerning her children. Ms. O'Hare was an Atheist, and her children were required to attend public schools where prayers were read on occasion. She won the cases, and public school prayer was banned. That was in the 1950's. Madelline and two of her children disappeared and were presumed murdered around 1991, and around a half million dollars worth of gold coins belonging to the organization went missing.

I had decided to write to them in 1980 because they would not likely see any reason to keep this from the public, it does rather strengthen the position of atheism, and they issued a magazine. Texas seemed to see flooding on a number of religious holidays in the '80s. It was certainly not my hope that this should remain a discovery known only to a few, or for that matter, something that was used for less than ideal purposes.

I know this is all anecdotal evidence but these are anomalous events, these coincidences. Scientific jurisprudence requires the investigation of any anomalous phenomena, does it not? In 1982 I learned of the terrible drought ongoing in Somalia and Ethiopia, and mailed a letter to Saudi Arabian Bechtel Engineering, A division of Bechtel Engineering, a major U.S. defense contractor.

Naturally I never got the daily weather in Saudi Arabia or surrounding area, but in 2003 or 2004 an article appeared in New Scientist magazine online stating that the Northern Sahara had been shrinking for the past twenty years or so, from Mauritania in the west all the way to Eritrea in the east, and all countries in between.

Did someone acquire a quantity of copper in Saudi Arabia and place it in a high location and leave it there for twenty years? Could a multi-billion dollar engineering firm afford to and be capable of doing this? I guess the drought in Somalia and Ethiopia continued a while longer,which are a little south of Saudi Arabia, not west of there were precipitation appears to have increased for twenty years, so it was a miss when it came to that.

Actually, about ten days or two weeks after I had written to Saudi Arabia, while watching the Weather Channel, the current weather in the Southern Atlantic was shown where a hurricane was tracking almost due east across the Atlantic Ocean straight for the Atlas Mountains in Western Africa.

That winter, 1982-1983, or the following winter,1983-1984 I forget which, I learned of a drought ongoing in Sydney, Australia, and wrote a letter to the American Embassy in Sydney with a brief description of what they might do in the circumstances they were finding themselves. Summer in Australia is during winter her in the U.S. Two weeks later I saw a small article in a newspaper about a storm that had hit Sydney, where it had rained "cats and dogs" for 24 hours straight. So, once again, it looked to me as though someone in Sydney did do as I suggested.

In the summer of 1988 a drought was ongoing in the Midwestern United States. Barges were running aground on the Mississippi river, soybeans and corn were shriveling in the heat. I wrote a letter to "Successful Farming" magazine detailing the remedy for drought, and a week or ten days later intense storms hit the Midwest with Des Moines, Iowa being among those hit hardest by flooding, that city being the location of the headquarters of that magazine.

Did someone perhaps endeavor to see if what I suggested really worked? Some months went by after the letter to Successful Farming during which I went to the library to read the latest edition of the magazine to see if anything was written concerning this idea that had been forwarded to them and the recent weather there, and eventually gave up.

I went through this experience one more time, in 1992. I moved to Las Vegas, Nevada in August of 1991. On December 30, 1991, after seeing something on television about the southwestern United States being in the grips of a seven year drought, I wrote and mailed a letter to Rolling Stone magazine on Wilshire Boulevard in Los Angeles, California. Therein I succinctly described how to implement a quantity of copper in a high location. January 6 or 7 saw precipitation across a wide area of the southwest. I was relieved when no flooding occurred, and began visiting the library to look through Rolling Stone magazine, expecting they would be progressive enough to handle the story. After several months of no article about anything remotely connected to weather modification, I stopped looking.

Five letters where it looked like someone may have acted upon my suggestion none of whom replied in any way. I e-mailed some people who were involved with trying to reduce smog in Mexico City and were using some kind of tall poles of copper or something and electric current to induce the particulates to condense out, which was the subject of another article in New Scientist I happened to look at when I saw the one about the changes in the Sahara, around 2003 or 2004, and I do believe Mexico City has been rainy pretty much ever since, but I really don't know if that is a major change for Mexico City or not, and certainly no one replied to me. Plausable deniability is the operative phrase,it applies to just about everyone.

How many indications that something unaccounted for has happened is that now, if we count Tesla and Hatfield, and throw all the strange weather in that has happened since 1900, all blamed on global warming, but possibly caused by curiosity killing the cat. The United States had a record number of hurricanes in 2005, including a good number that made landfall, and the following 2 years saw almost no hurricanes in the vicinity of the U.S.

Does Mother Nature change her mind that abruptly, or did someone place lead somewhere along the southeast coast to ward off hurricanes in 2006 and 2007, something that obviously wasn't done in 2005? There was a drought in the Southeastern United States in 2007 with water levels falling dangerously low. Could that drought be the outcome of lead being placed along the coast to prevent hurricanes from approaching?

Additionally, I have conducted what experiments I could on my own. From May of 2002 until probably early 2008 there were half a dozen objects resembling circular bird cages about 6 feet high each and 2 feet in diameter, all made of nearly pure copper, sitting on a hillside in the desert southwest of the U.S. The total weight of this row of copper tubing was between 300 and 350 pounds. The height of the hill upon which it was resting, on the southwestern side of this hill, is only 1500 feet or so above the valley stretching out to the south and west, so it had a small unobstructed space before mountains 6 or 8 miles distant block the signal from the copper.

The quantity of copper needs to be greater, and the elevation of its location needs to be higher, to see more robust precipitation. Nevertheless there were observable effects, including numerous small thunderclouds dropping rain in small areas surrounding the hill with the copper. Of the 10 or 12 times I have driven past the location within a few miles, small rain clouds covering just a few hundred yards have splattered my windshield about 75 % of the time. It was not unusual to be hit by three or four of these small storms 10 miles either side of the emplacement. There was quite often a persistent narrow band of clouds directly over this location,with the trail of clouds only a few hundred yards or so wide stretching 3 or 4 miles in either direction from the hillside where the copper rested.

When I first hauled the materials up the hillside to where they rested,I laid them all on a flat ledge horizontally and didn't return for two weeks, at which time the wind in the area was quite excessive. Once they were tied together with copper wire to straight bars of copper tubing,and standing vertically, cloud development became much more frequent in the area, and the winds settled down. On Labor Day around 2004,I and two partners in mischief ascended to the copper with more copper and assembled the sixth bird cage object, carried all six up the hill a little further, and began our descent at 4P.M. that Monday afternoon. Rain began at 4P.M. that Thursday over a wide area including where I live, and since there is an hour time difference between where I live and where the copper was, the elapsed time was closer to 73 hours.

Actually, as this writing continues, I made a journey to the sight where the half dozen copper devices were left, with a friend, who also happens to own the property, with the intention of taking some pictures to add to this book. After climbing around the entire hill, we found where the site had been, but found only a hack saw, and one of the copper wires that we had used to tie one of the copper constructions to a nearby rock to keep it from being blown over by the wind, with the loop that had gone around the rock still tied.

So the experiment came to an inglorious end when someone noticed the strange looking objects while hiking through the area, looked closer, decided that it was something valuable, and stole it. It must have come as a shock to whoever found those things, to see a half dozen odd looking contraptions in the middle of nothing but rocks, sand, cacti, and sagebrush. I do have witnesses to the experiments though. The theft we think took place in early to mid 2008. If one were to look at the weather in Kingman AZ from 2002 through 2007, it might just show above average rainfall. The

location was about ten miles toward Las Vegas
from Kingman.

Chapter 8: From The Beginning

-------------------------------------------------
" That which stumbles at first can
become sleek as a roadrunner."
JB
-------------------------------------------------


My interest in copper and the atmosphere began
with a book titled "Modern Physics and
Antiphysics"[8], which I read back in the 70's. I
became fascinated by the part where the author
discusses "The Absolute Frame Of Reference", and
the search for it. What we can distinguish about
things and their movement in space always comes
from a frame of reference that is relative to
another observer, or relative to some stationary
object. Though that stationary object is moving, it
is moving right along with an observer, and is thus
apparently motionless.

So our frame of reference is our own Solar
System, since all within it is in orbit around the
Sun, and following the Sun as it orbits the Milky
Way. Knowing the absolute frame of reference
would enable one to distinguish between one
square foot of space and it's adjacent square foot
of empty space. Areas of space are things that are
not labeled in any way, are completely identical,
and consist of nothing.

If one could account for all motions that the Earth is undergoing, and travel backwards at the exact speeds and times, you will have reached a point of absolute motionlessness,with the exception of your bodily functions, but who knows? Maybe something weird would happen then, when you are no longer moving in space. There has been speculation about zero point energy that ties in with the absolute frame of reference and knowing one's exact speed and direction as well. Anyway,the discussion began with Michelson and Morley, and the search for the "ether wind". I think that was the title of the chapter.

Michelson and Morley were searching for an ether wind, a kind of stratum in empty space through which things pass, theorized to exist to explain gravity, or the passage of light through vast reaches of space, or Newton's Laws, or it involved Einstein's discussions of how one observer, being in a different place from another observer, may see an event in an entirely different frame of reference as actually occurring differently from the account of the other observer.

Anyway, there were various hypotheses about the subject matter, of the void, and was anything in it. Also how,indeed,does one distinguish between one square foot of empty space and another, since we could be sliding any which way between any astounding number of empty square feet,and that thus far we have absolutely no way to distinguish one from another.

Actually, the Solar System, in orbit around the Milky Way, is moving somewhere around 500 thousand miles an hour. One could be to the moon and back in an hour at that speed. So the problem defined,as time went on an idea came to me about how one might possibly conduct an experiment to find the absolute frame of reference. The problem remains unsolved to this day, and many think it insoluble. Obviously I didn't solve it either.

First of all, if one were to conduct such an experiment, the experiment would take place in a fixed location on the surface of the Earth at some specific time. Each day, wherever we are going makes a full revolution from our perspective on the surface of the Earth. Though the Earth continues in one basic direction, at noon it is 180 degrees different from whatever direction and speed it is going at midnight,to an observer in a fixed location on the Earth. For example, suppose at noon the direction of the Earth is straight up over our heads. At midnight it would be straight down beneath our feet. This is the most rapid change.

Starting with the big bang, all of the things that eventually become our Solar System are thrown out at enormous speeds along with everything else. We still retain some inertia form that event, so that is one motion, if we assume that the collision of the two super massive black holes that created our known universe, or whatever theory about it coming out of a singularity, was absolutely motionless at the time.

The Big Bang could have been sliding one direction or another while it happened, so determining the ultimate direction of things mathematically might ultimately prove impossible, since one could never have the figure for other possible motions. Anyway,after the Big Bang,there is galaxy movement within the cluster of galaxies that includes the Milky Way, our rotation as a star system with planets around our galaxy,our rotation around the sun, and the Earth's rotation on it's axis.

Given that there are 5 or more different motions to calculate, and high speed velocities, the curving path would be as near a straight line as one could get for any short distance. The fact that it is not perfectly straight would not reveal itself for thousands of miles. That consideration understood, one could conceivably find where exactly the matter in our solar system is going by accident. Lined up in the nearly straight line that all the local matter is traveling in, there could be static electricity left in our wake and that could eventually be detected and collected.

This idea tied in with the discovery withheld from the world in the novel Atlas Shrugged. The thing that intrigued me about that book was the invention. The invention consisted of some device that captured ambient static electricity and converted it into electricity. The ambient static electricity would be left in our wake, was my conjecture, as we sped through the universe at incredible velocity, perhaps.

I even hypothesized at the time that since space is currently indistinguishable, in so far as telling which identical area of space is which, and that the Big Bang being such a colossal explosion, perhaps the matter flying apart exceeded the speed of light, and could even be doing so now, since the universe seems to be expanding very quickly.

For all we know our limited viewpoint wouldn't tell us exactly how fast we are going anymore than where or in what direction. So my thinking went, what if one knew the absolute frame of reference and had just the right type of arrangement which would be easy to make with a length of copper refrigerator tubing, aligned so that as it flew through space exactly so, it was in line with our actual direction through space and would be in position to capture the static electricity? Maybe that was how it could work.

The ambient static electricity was the main thought that inspired me to eventually try the experiment, though finding the absolute frame of reference might be necessary to actually be in position to capture static electricity and turn it into electrical current, or so my conjecture went.

So the plan is to have an almost straight sort of arrangement or another of copper and hope by sheer luck, by trying a number of different angles and directions with the thing, something unexpected happens if it happens by chance to be at or near the direction we are ultimately racing along at, the actual direction of all the matter in the vicinity.

I thought that would be the only way to get an inkling of this absolute frame of reference, by deliberately stumbling upon it by giving this copper thing enough different tries. What was the rush, anyway, I probably would never find it, but even if I didn't I would still have the copper, and could continue with it, maybe refine the experiment with knowledge of the direction, say, of the solar system through the galaxy at 12 noon in my neck of the woods, get a voltmeter, etc. That was how it all began.

So having determined it would probably take years to get anywhere, but impulsive enough to try anyway, I acquired two continuous lengths of circular copper tubing that could be stretched out to a length of twenty feet or so, each loop of coil about a foot apart. My main conjecture was that ambient static electricity might accumulate in some circular tubing just because copper conducts electricity so well,something that could possibly be captured better along the length of the copper depending on how it is positioned, so I stretched these two out the length of the upstairs bedroom, and at the lower end was a car battery out of power,and the upper ends were by the one window which I opened to the north, and the one window I opened to the west.

The absolute frame of reference was not uppermost in my mind as this thing started. Ambient static electricity was out there, and the possibility of capturing it was what I patiently began to explore.

I think I noted the time to myself, and three days go by, and there is this storm moving in to the neighborhood. So,being fascinated by the weather, I went out in the back yard, and was watching the lightning, and listening to the thunder, and the storm was coming in close, the thunder was getting louder, and heard in less time after the flash, when suddenly the loudest clap of thunder I have ever heard shook me to my roots, and in that instant, the translation of the name Ayn Rand into the phonetic pronunciation "Aye 'n Rained" like an Irish person talking or something struck me,and I finally perceived what had eluded me up to that point. So my mind raced with the possibilities. There was the start of my realization that weather modification might be a possibility, and had probably long been known, and withheld from the public.

So, as it turns out, it seems it was a dumb experiment that lasted just three days with no chance of succeeding unless some sensitive measuring instruments were used but I did it, and that blast of thunder sent me immediately away from ambient static electricity and the absolute frame of reference to weather modification and collecting electricity from lightning.

There I was oblivious to the possibility that the copper tubing by those two open windows might have had something to do with the fact that this storm was bearing right down on us from the northwest, I had forgotten all about them, when that clap of thunder and the thunder bolt of realizations in my head happened at the same instant. It was plain that the copper must have been causing a path of least resistance which the incredibly numerous atmospheric particles,being small, and having electrons, would follow.

Actually it may have had some short term impact on the storm in the area, maybe enough to bring one lightning strike quite near, but it was certainly too small an amount to actually generate the storm, at least that is what I thought.

So the actual beginning of all this began with an idea about something else entirely, and the first events where I experienced experiments with copper involving the atmosphere were quite inconclusive. I think that by now,though,enough of what I have pointed out convincingly supports the theory, and the reader can see that there is really something to it. We would all like to know if there is really something to it,I hope, by this point in my presentation.

But,really,to properly conduct a hot or miss type of experiment such as I had begun to try to carry out and quickly abandoned, one would need a huge room like an aircraft hangar or a large pole barn, a stand to place super conducting material on that one could turn in any direction with control mechanisms,a one hundred foot long or more super conducting composite material in a straight circular tube, a plethora of sensing equipment all along the length of the super conducting material,and a bank of computers to monitor the whole event, all the while calculating exact geometric position in the universe, motion of Earth's axis, orbit around the sun, and the orbit of our solar system around the galaxy.

Then one would program the thing to move position every two minutes, or some such small time, and let the thing run itself for several months, return, analyze the data, find no anomalies, run it another six months, and maybe find something odd occurred during one two minute stop, bring up all the sensor data recordings from the event, calculate the exact position, the time it occurred, the day of the year, etc.

With this data repeat the experiment each day with minor variations for three weeks until, who knows? One day unusual readings occur again. Maybe one could detect where everything is going and at approximately what speed eventually. And for all we know electricity could be there for the capture. Probably not though, I'm just trying to defend the idea. Besides, discovering the Absolute Frame of Reference would be a terrific discovery, and stumbling upon that accidentally on purpose seemed about the only way to proceed to find it.

After all,the beginning of all this goes all the way back to Tesla,and it turns out that 'tis I who, apparently, has carried on as Tesla predicted someone would eventually. The invention depicted in "Atlas Shrugged" was one more stop along the way in accumulating information and possible insights into what eventually became what one could call "The Primordial Speck Theory",if it were to actually begin to be considered seriously, and if one Cosmologist were to refer to it for whatever reason, that name would work as well as any other.

If other Cosmologists were also familiar with this theory, and I certainly hope that happens, they would immediately understand which theory was meant. But it is plain that I have discovered nothing that hadn't been discovered much earlier by other humans, and what I have done is simply uncover what had apparently happened in the past and interpret what nature is up to, perhaps better than what the laws of physics of today and official meteorological dogma preach, with a few new ideas in Cosmology.

Soon after the blast of thunder in June,1980 when I first realized these new possibilities I soon brought those copper coils outside to try to see what, if any, effect they might have, and these were dumb experiments too, because I think I had only 40 pounds of copper, and no real high place to put them. Even so,I think they may have contributed to cloud development,and when I placed them outside cloud development did occur, and seem to peak in intensity around 72 hours after putting them somewhere, and usually look as though rain could come,but usually little or none. I did add some to it, over the early 80's, to where I was carrying around about 75 pounds of copper. I usually only ventured out when forecasts called for cloudless skies.

All I can conclude now is that nothing I did at that time actually proved anything definitively, but eventually I came to realize a path of least resistance really does seem to be at the heart of the matter, one involving the primordial speck, and not N2 and O2, although the nitrogen and oxygen are involved via gravity. So I realized it would need quite a bit more copper than what I was experimenting with back then in the early 80's to actually produce a decent storm system.

I had seen what had happened in Texas, Australia, and Iowa where drought ended suddenly shortly after my letters were sent there. After a while I concluded those people had probably tried a ton of copper on some hillside, maybe even two. Copper was around $1.50 a pound back then. Two tons of copper cost $6000 back then, petty cash for a lot of wealthy corporations and people. I never advised anyone to try any more than 500 pounds, at least on the first try, and all these events got me eventually contemplating further what had probably happened in those places.

Whenever an amount of copper in a high location exceeds a half ton or so, hydrogen nucleo-synthesis in the atmosphere due to the intense concentrations of primordial specks probably escalates to the point where hydrogen pairing and meeting single oxygen isotopes produces incredible volumes of water in the seventy two hour incubation period in which the storm front is developing. When the precipitation arrives it is too much. The more copper there is, the greater the intensity of the electromagnetic waves that propagate outward from them. The atmospheric components would be pulled that much more vigorously along by an even greater concentration of primordial specks,and the range that the effects would reach would grow larger as more copper is used. So, all the theorized effects would increase or decrease in correspondence with how much copper is used.

So, in an attempt to find an explanation for events meteorological that shouldn't be happening according to the laws of physics as they are understood now, but that I was convinced were happening nonetheless,this theory has grown,and I think it fits quite well in the overall scheme of things. As time goes on,more evidence may turn up to shed further light on some of the cosmological questions, I think we all know the meteorological questions are all but answered.

Chapter 9: Suppositions and Predictions

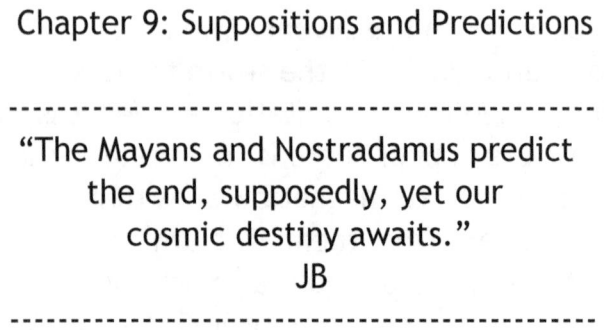

------------------------------------------------
"The Mayans and Nostradamus predict
the end, supposedly, yet our
cosmic destiny awaits."
JB
------------------------------------------------

Last but not least, a firm record of the entire
Primordial Speck Theory needs to be drawn up,
with all its suppositions and predictions. That way,
we can all see in the fullness of Time whether the
predictions are fulfilled, and if the suppositions it
makes hold true. Someday perhaps more sensitive
detecting equipment will be designed that will
distinguish three or four different particles in
random space that previous to this time were
below the threshold of detection.

If that happens, a one particle hydrogen
precursor theory would obviously be wrong,and not
stand the test of time. Actually, though, most of
the suppositions made by the theory would all
probably still hold even if the number of extremely
small things was more than one type. The different
types would combine in whatever way to become
hydrogen whatever the dark matter and dark
energy are, and we are currently in a position
where only theory will shed any light. For the sake
of simplicity the theory goes with one fundamental

particle,since the most pulverized thing possible doesn't seem likely to have choices.

Supposition 1: That undetected dark matter and dark energy in the known universe are primordial specks, precursors to hydrogen made of combinations of matter and energy too small to detect, probably one fundamental particle in astounding abundance.

Supposition 2: That these primordial specks are most likely magnetic monopoles. A cascade of these begins when primordial specks are in sufficient concentrations resulting in nucleo-synthesis of hydrogen through pairing up and adding of additional monopoles in multiple steps. The likelihood is that we will never know how small each individual primordial speck is, nor the steps it takes to unite with others of its kind and become a hydrogen atom, comprised of a proton with three quarks within, and an electron in orbit around it.

Supposition 3: Being magnetic monopoles, the primordial specks are unstable, energetic, and spinning at varying speeds. They will accumulate along a path of least resistance such as can be created by placing a quarter ton of copper as earlier specified near a mountain top facing the prevailing westerlies.

Supposition 4: The accumulation of primordial specks in increased concentrations along a path of least resistance and the resultant nucleo-synthesis of hydrogen in the atmosphere will result in newly formed hydrogen atoms pairing up as stable H2 isotopes, rising to the ozone layer, joining with a free oxygen atom there, and in great quantities becoming water molecules that join developing clouds.

Supposition 5: The primordial specks gathering in increasing numbers along the path of least resistance,all traveling in the same general direction, will gravitationally pull the electrically inert N2 and O2 that accounts for 99% of the known air along with all of them along the path of least resistance. The other 1% of the known air, argon, methane, neon, carbon dioxide, etc., will also succumb to the gravitational pull of the mass of the primordial specks, and join the schooling cascade. All this, of course, results in falling barometric pressure, cloud accumulation and ultimately, precipitation, usually around 72 hours into the experiment. Supposition 4 also figures into the clouds and precipitation.

Supposition 6: That the combined impact of suppositions 4 and 5 along with perhaps some other additional factors helping such as the Jet Stream, static electricity, magnetic field changes, etc., proves that a path of least resistance can be created with a row of copper tubing in a high location, resulting, if the experiment is properly conducted and uncontaminated, in precipitation occurring 72 hours, give or take 8 hours, into the experiment. Also that cloud development will start promptly within an hour or two of a sudden insertion of copper of sufficient size near a mountain top, and build in intensity until the release occurs in the form of precipitation.

Supposition 7: That since we have established that dark energy and dark matter is in the form of particulates that are a combination of both matter and energy, the excess of the dark energy over dark matter can be explained by the 52% excess dark energy consisting of the gravitational pull of massive objects further away from us than the known universe. This gives us a 90 to 92 percent percentage of everything in the known universe being primordial specks  and 8 to 10 percent being elements, hydrogen on up the table of elements, along with a few other stand alone particles like neutrinos, and released energy from matter in various forms.

The increased acceleration of the known universe is explained by this and it is predicted that other objects of mass will be discovered outside the known universe within the next 50 years. Right now the idea that 52% of the dark energy is in the form of distant objects converts the dark energy into things, then, that are a combination of matter and energy, just like here, it is simply that we are getting no information from them other than their gravitational pull. The mass of the hypothesized more distant objects have never figured in to calculations because we have no way currently of finding it out. We only have the calculated strength of a force trying to pull the known universe apart.

Supposition 8: That Space has not, as is thought by many, been shown to admit of alteration in any way. That photons, a form of energy with no mass, should be bent by massive objects in their vicinity, proves nothing other than that the photons changed course, not that the Space changed. A more thorough theory of gravity with space as a fabric thrown out will emerge from this within 20 years. Space is probably absolute,eternal,and it would be hard to remove nothingness from somewhere.

Supposition 9: Particles do not pop out of zero point energy or some such things, they coalesce from smaller particles in some sort of cascade of joining and combining. These primordial specks will eventually be concluded to exist, and be regarded as the most abundant thing in the universe by almost all scientists and literate people in 40 years.

Supposition 10: That given the considerations of Supposition 7, the known universe is not the entire universe. Given that, our known universe need not collapse back in on itself, and the accelerated expansion of the universe will probably continue, and eventually the cold, lifeless things that were our solar system will be absorbed by a super massive black hole, which will eventually collide with another, spawning another 100 billion or so galaxies, so the entire universe will continue to exist, and we will never know its full extent.

Supposition 11: That an experiment with copper in a high location, yielding increased concentrations of primordial specks would show more hydrogen atoms in random air samples taken 72 into the experiment than an experiment of the same type that instead used a row of lead sheets positioned from north to south facing the prevailing westerlies, with random air samples taken 72 hours into the lead type experiment of identical size as the previous experiment.

For successful tests of that prediction, however, one must realize that in the copper part of the experiment, the barometric pressure will in all likelihood be much lower, the relative humidity much higher, and it will probably be raining. Thus, some of the hydrogen found at that moment might just have come from a water molecule breaking apart, so one would have to factor in an estimated quantity of those. I really don't know if water molecules break apart easily or not. One would also wonder how many water molecules have brand new H2 isotopes, or half new, half old, but determining that is not possible.

It is predicted that numerous experiments of the type mentioned will be carried out, and confirmation that additional hydrogen is present in greater parts per billion in the copper part of the experiment beyond an amount correctly calculated for water molecules breaking apart will happen within 30 years. Based on the results and the amounts of additional hydrogen being accurately calculated, a computer simulation will, within 30 years, be able to estimate fairly accurately the additional hydrogen being created over a specified time within the bowels of the Earth, within the Sun and the Moon, and any other object in the heavens within range of telescopes.

Supposition12: That the dangerous storms that brew out over the oceans and are known by the names hurricane, typhoon, and cyclone could be wholly eliminated if mankind cooperated with each other and diverted a substantial portion of the water vapor and static electricity out over the oceans towards land. Additional drainage of the

Oceans could be accomplished by positioning more than the usual amount of copper near the north and south poles. This last mentioned activity would also rebuild the polar ice caps. The prediction is a sharp falling off of the aforementioned storms in the next two decades.

Supposition 13: That tornados could be eliminated to somewhere within a 98% reduction, year to year, with a little experimentation with different ways of creating a path of least resistance. It is predicted that the prevalence of tornados will fall off sharply within two decades. Flooding and drought also become much less troublesome in the next two decades. It is predicted that a barometric pressure hotline will come into being some time in the next two decades. It is also predicted that this hotline will no longer exist in 50 years, along with the weather channel.

Suppositions 14: That the primordial speck will never be detected,but that new sensing equipment designed to detect a hydrogen particle materializing out of seemingly nowhere will be developed within 30 years and be successful in finding such events.

Supposition 15: Some time in the last three decades of the twenty first century, a mathematician will calculate accurately the smallest magnetic monopole the universe can possibly make, and computer simulations with the new data inputted for the mass of the primordial specks will show a high degree of accuracy in predicting actual conditions.

Supposition 16: That the insurance industry and the various branches of government in the United States will revise the term "Acts of God" to "Acts of Nature With Possible Human Involvement", or some other more accurate term when referring to weather events by the end of the 21$^{st}$ century.

Supposition 17: That since we have restored Space to her former glory, and space/time being a mistaken idea,or something applicable to equations but not reality, then the Big Bang didn't occur as theorists now maintain it did with space/time collapsing into a point together with all the mass of the entire universe contained within, but was rather likely the collision of two super massive black holes, in a very empty quadrant of the much larger universe, since these two super massive black holes that collided vacuumed up all the matter in the area to a distance of 10 billion light years in every direction before colliding. It is predicted that Einstein's theories will still hold, with the new condition that time is variable, but space is not.

Supposition 18: That computer simulations with the suppositions listed above will likely yield an accurate picture of reality, and thus be able to successfully predict atmospheric conditions 72 hours into the future within a small variable, since no two storms are exactly alike, and thus complex systems such as the atmosphere on Earth can have successful predictions made about them, despite whatever chaos theory may say to the contrary.

Supposition19: Global warming will, within 30 years, no longer be viewed as an imminent threat. Global temperatures stabilize within that time and begin to show little variation from year to year, with ocean levels well below the highs of thirty years previously.

Supposition20: The increased availability of water on land, and improved runoff from rivers to oceans begins to slow the extinction of species considerably in 20 years. The plant and animal life in the wild both on land and in oceans and fresh water habitats begins to show marked increases in populations in 30 years.

Supposition21: Desalinization plants will no longer be planned in twenty years, and in the case of those already existing,some are decommissioned and converted to sports stadiums within twenty years.

Supposition22: Within fifty years, an attempt to harvest electricity from lightning will have been made,and based upon actual data from the ongoing project in a remote ocean location, one such type electricity harvesting rig in each of the major oceans is eventually built, by the end of the century.

These predictions are all predicated upon the idea that this book gets published, and the estimated times begin at the date of the first publication of this book. Even if the ideas presented here only reach a few people in the first few years after publication, eventually some interest should get generated, and two or three years shouldn't affect the predictions much.

Maybe a lot depends on the attitudes of people toward tampering with nature. The only way that attitudes can change and more and more people envision a brighter future involving a little nature tampering is if books like this are written and read. I certainly got nowhere with the meteorological community by sending a one page description of this theory.

## CONCLUSION

The problem couldn't be more plain. This feature of nature that I've tried to describe and talk up needs a full investigation by the scientific community. I watched the presentation about Ardi, the 4.4 million year old hominid fossil found in the highlands of Ethiopia, and all the nations and scientists who collaborated in the excavation and examination of the fossils, and it seems that a good deal of time and trouble was spent on it.

It just seems a shame that a small fraction as much time and trouble couldn't be allocated to furthering human knowledge in the investigation of the idea proposed here. Just a ten minute segment on Mythbusters would do it. A little explanation, the experiment, some time lapse photography, and there we go.

Still, a stint on Mythbusters that appeared to support the assertions made in this presentation wouldn't constitute scientific proof either, unless enough experiments were tried that coincidence was completely ruled out, and all the repeat experiments wouldn't make for interesting entertainment. But that 10 minutes of affirmation might be enough to persuade the public to favor the idea that the scientific community should go through all the necessary procedures required to actually prove it, or disprove it, as the case may be.

As mentioned earlier, people like to know things. I do believe there is something to what I have tried to present. The invisible forces of electromagnetism and gravity are all around us continually; that the two could combine to produce effects along with the hypothesized primordial specks such as described makes sense. It fits all the data. Veni, Vedi, Vici. The  seeing is conceptual in the case of primordial specks, and the forces of nature.

No other explanation or theory fits all the facts as well. As for how dangerous this kind of weather modifying activity could be,it doesn't get any more or less dangerous based on how many people on Earth know of it, it is an inescapable feature of having a tool capable of doing as described earlier. Knowing always involves an empowerment, and, surely, once enough people become aware of this and begin to discuss it, our birth as weather modifying beings will have begun, and before long, proper management of fresh water on a world wide basis will come easily and safely.

We have a low-tech tool with a wide range of uses, some of which were mentioned earlier. There is some work involved implementing the tool, some equipment, but surely possible almost anywhere on the globe. The questions each reader must ask themselves are: Does what Mother Nature does without human intervention have such divinity that they would prefer to suffer whatever that may be including all the bad types of weather that we have already discussed,or would the reader rather not experience drought,flooding,tornados, hurricanes,etc.,and would the reader like to know the truth and be the wiser or prefer to remain deluded?

The question also remains whether weather events in the future will be done by nature alone, or with some assistance,and that question won't be easy to answer in a world where we don't learn this. That world will be a world where this feature of nature will still exist, whether acknowledged or not. The beauty of nature stands a chance only if weather is optimized for the wild inhabitants. Traveling through Yellowstone National Park about 5 years ago we saw a Yellowstone with huge tracts of scorched burnt trees, from forest fires not long before. Would that have happened if man had been actively intervening with the proposed "new" technology when forest fires occurred? Was my appreciation of nature enhanced by the destruction caused by unrestrained mother nature? Of course firefighters and equipment, planes and helicopters dumping water and chemicals were used to fight the fires, but is that the most effective way to combat a forest fire?

Any rational being, having read what arguments have been put forth here, would conclude that it is our duty in the natural scheme of things to perform custodial work on the planet we live on. Had humans not arisen as the apex creature on this planet, some other animal would have eventually, and so, in the cosmic scheme of things, a creature at the top of the food chain will evolve, and eventually fill the position of the only creature on the planet capable of modifying the weather. Does the universe expect that creature to carry out his duty? No, the universe just makes it possible.

\* References

1. Tesla: Man Out Of Time
   By Margaret Cheney   C.1981
2. Tesla: Man Out Of Time
   By Margaret Cheney   C.1981
3. Dark Cosmos; In Search Of Our Universe's
   Missing Mass And Energy
   By Dan Hooper        C.2006
4. Dark Cosmos; In Search Of Our Universe's
   Missing Mass And Energy
   By Dan Hooper        C.2006
5. Dark Cosmos; In Search Of Our Universe's
   Missing Mass And Energy
   By Dan Hooper        C.2006
6. Weather Modification By Cloud Seeding
   By Arnett S. Dennis   C.1980
   Weather And Climate Modification:
   Problems And Progress
   By Thomas F. Malone  C.1980
   Weather Modification: Prospects
   And Problems
   By Georg Breuer        C.1980
7. Atlas Shrugged
   By Ayn Rand           C.1957
8. Modern Physics and Antiphysics
   By Adolph Baker        C.1970